Robert B. Leuchars

A Practical Treatise on the Construction, Heating and Ventilation of

Hot-Houses

Robert B. Leuchars

A Practical Treatise on the Construction, Heating and Ventilation of Hot-Houses

ISBN/EAN: 9783337805302

Printed in Europe, USA, Canada, Australia, Japan

Cover: Foto ©berggeist007 / pixelio.de

More available books at **www.hansebooks.com**

A

PRACTICAL TREATISE

ON THE

CONSTRUCTION, HEATING, AND VENTILATION

OF

HOT-HOUSES;

INCLUDING

CONSERVATORIES, GREEN-HOUSES, GRAPERIES,

AND OTHER KINDS OF

HORTICULTURAL STRUCTURES.

WITH PRACTICAL DIRECTIONS FOR THEIR MANAGEMENT, IN
REGARD TO LIGHT, HEAT, AND AIR.

ILLUSTRATED WITH NUMEROUS ENGRAVINGS.

BY ROBERT B. LEUCHARS,
GARDEN ARCHITECT.

NEW YORK:

C. M. SAXTON, BARKER & CO., 25 PARK ROW.

SAN FRANCISCO: H. H. BANCROFT & CO.

1860.

Stereotyped by
HOBART & ROBBINS;
NEW ENGLAND TYPE AND STEREOTYPE FOUNDER
BOSTON.

TO

PROFESSOR SILLIMAN, JR.,

This Treatise,

DESIGNED TO PROMOTE THE ADVANCEMENT OF EXOTIC HORTICULTURE,

OF WHICH HE IS A ZEALOUS PATRON AND ADMIRER,

Is Respectfully Dedicated,

BY HIS OBLIGED AND OBEDIENT SERVANT,

THE AUTHOR.

PREFACE.

HAVING for many years past devoted my attention to the subjects treated of in this work, and from the general call for appliable information thereon, I have been induced to give it to the public, in the full persuasion that it will be acceptable to horticulturists, gardeners, and others engaged in this department of horticulture. From the numerous inquiries which I have received, there appears to be a great want of practical knowledge on these subjects; and though much information may be gleaned from various English works, they are either unobtainable, or the information is inapplicable to the wants of this country.

When I commenced this treatise, I intended it as a series of articles for periodical publication; but the development of the subjects, and the accumulation of facts, swelled it to such a size as to render its publication in that form impossible. In preparing it for the press, in its present form, I have been desirous to add nothing but what is necessary to a full understanding of the subject in hand, and have given figures and diagrams where illustration is required.

The changes which have occurred, during the last twenty years, in the method of constructing and managing horticultural structures, render the works of that period of little value to gardeners at the present day. I have here given all the latest improvements and most approved methods at present in use, with plans and suggestions for their further improvement.

From what has been said, I hope no one will suppose that this treatise is given as a complete work on Exotic Horticulture. Much has yet to be learned, on many points connected with hot-houses, which futurity will, no doubt, unfold.

My warmest expressions of thanks are due to Professor Dana, of Yale College, for the generous manner in which he has favored me with his opinions. The readiness with which that gentleman has replied to my inquiries, on matters of science relating to my subject, even in the midst of his laborious literary pursuits, shows how willing he is to aid the most humble inquirer. This expression of thanks is due from me here, as the only way in which I can sufficiently show the high value at which I estimate his kindness and liberality.

R. B L.

Boston, Oct. 3, 1850.

INTRODUCTION.

THE object of the following treatise is chiefly to lay before its readers a series of facts and observations relating to the construction and general management of all kinds of horticultural structures, drawn from the developments of science, and an extended experience, with the view of leading those who are interested in this delightful pursuit to a more practical inquiry regarding the comparative cost and economy of the various methods now commonly adopted, as well as to draw the attention of practical gardeners to the utility of studying the theory as well as the practice of those manifold operations on which the success of exotic horticulture depends.

In a short treatise, on such comprehensive and varied subjects, it is impossible to be strictly scientific; but we have endeavored to show the rationale of those methods and operations which we have here recommended, and which have been successfully carried out by us in practice. The treatise is avowedly a practical one, and intended chiefly for the use of practical gardeners, and those desirous of obtaining that knowledge which is necessary to enable them to superintend the erection and future management of their own garden structures. In the management of hot-houses, there is a systematic regularity required in all the operations, a neglect of which is generally attended with disorder and confusion. In fact, there is a system, the details of which succeed each other like the links of a chain, each operation being essentially connected with the one immediately following and preceding it; and here we have a most encouraging truth, that the more scientific our principles of working, the more simple and easily performed are our operations, and the more reliable are the results.

It is doubtful if any branch of horticulture has received less aid from science than that which forms the subject of the present work. Science has indeed been brought to bear upon horticultural generalities, but, as far as regards its application to exotic horticultural details, it is little better than a sealed book; and hence it is that we find cultivators clinging to antiquated systems, which the plain demonstrations of science and practice are daily proving to be absurd. Amateurs, who adopt exotic horticulture as an amusement, and pursue it with enthusiasm, are very apt to be misled by the advice of those who are more ignorant than themselves. They are easily led into extremes; and nothing is more common than for such persons, in their zeal, to adopt one error, under the plausible pretext of avoiding another.

From the importance of LIGHT, HEAT, and AIR, in the economy of vegetable life, it is obvious why an architect is professionally incapable of constructing a house for the growth of plants or exotic fruits, without possessing a knowledge of the requirements and functions to be performed by the silent inhabitants; and hence it is, that, by studying these principles in connection with other branches of science, we arrive at the end more rapidly and successfully. In other words, cultivation becomes more certain as it becomes more scientific. Practical illustrations will hereafter be given, to show that horticultural structures, instead of being subordinate to architectural arrangements, as they generally are, must be accommodated to the necessities and requirements of vegetable life, before satisfaction can be afforded to the possessor, or cultivation carried on in perfection.

If we take a glance at the progress of horticulture in Europe during the last twenty years, we cannot fail to perceive that its advancement has been parallel with the developments of chemical and physiological science. Almost every succeeding year has brought with it some new and important improvement in practice, and thrown additional light upon some hitherto disputed question. Although gardening has, in some solitary instances, been remaining stationary, the cause is by no means obscure. Gardening is encouraged just in proportion to the satisfaction it affords. It gives satisfaction according to its success, and success is in proportion to the amount of practical experience founded upon a scientific basis.

Whatever causes exist to prevent the operations of gardening from being carried out on scientific principles, it is nevertheless true, that no methods can be generally applicable, or universal in their results, that have not such principles for their bases. To be guided by them, it is not necessary that the gardener should be a mere reader of books, a studier of theories, or a continual performer of experiments; he must add to the precepts of others the acquisitions of his own experience, and aim constantly at progress, by learning practically the principles upon which his operations rest for their success. It is not the lot of every one to discover truths hitherto unknown, but almost every one engaged in the practice of horticulture can do something towards improvement, by enforcing those already known by stronger evidence, facilitating them by a clearer method, and elucidating them by brighter illustrations.

There is a wide-spread antipathy to all kinds of book instruction, and book gardening is ridiculed by many who call themselves gardeners. It is, nevertheless, a well ascertained fact, that those who rail at book practice are not only the worst practicals, but also the worst theorists, and the worst reasoners upon matters of practical import. Indeed, there are few who are more slow to recognize the benefits of the valuable knowledge to be found in the works of the eminent horticulturists of this country, than those by whom it is most required.

Notwithstanding the valuable works which have lately been given to the world, on horticulture and the kindred arts, by eminent writers, little or nothing has been done in the department embraced by this treatise. Horticultural structures of all kinds continue to be made, and managed, with the same disregard to the actual habits and requirements of plants, as they were a century ago. And, though some structures of this kind have been constructed upon plans and principles in accordance with modern knowledge, yet these are a very small exception. Many apparently fine structures could be pointed to, which are rendered comparatively useless for the purposes for which they were built, on account of a deficiency of knowledge on the part of those who superintended their erection.

It is a common error for gardeners, and others, who erect glazed structures, to suppose that the kind of house perfectly suitable in one place will be equally so in another; or that the same arrangement answerable for one purpose will answer equally well for all purposes to which a glazed structure may be applied. Some of the consequences of these errors will be more particularly specified in a subsequent part of this work; as also the external forms and internal arrangements which we have found most suitable to the different purposes. The influence which a servile adherence to old methods has upon the progress of horticulture, is chiefly manifest to those who are most liable to be censured for innovations. Yet it is doubtful whether the odium incurred is not more than compensated by the pleasure which arises from the rewards of perseverance, by which we are enabled to abandon bad systems, as we gain more confidence in those that are better.

It is said that practice is the best of all teachers; that as our practice is lengthened, our experience is increased. However this axiom may hold good in the common affairs of life, it is frequently reversed among practical men, and years pass away without any enlargement of knowledge, or rectification of judgment. There are, indeed, many who never endeavor to improve, notwithstanding the opportunities which may be afforded them. The opinions they have received, and the practice they have learned, are seldom recalled for examination, and, having once supposed them to be right, they can never discover them to be erroneous. From this preconceived acquiescence, few are entirely free; from a dislike to apparently superfluous labor, and from a fear of uncertain results, many stand still when they might go forward.

Some may say, that if a practical man performs the operations which others have taught him, and succeeds as well as others have done, he does all that can be expected from him. But this is doing nothing for improvement, and very little for himself. It is every man's duty to endeavor to excel, both on account of his profession and of himself, as well as those who employ him. It is easy to perceive that a gardener must not only know how to do, but have his reasons for doing. A man who continues to do his annual operations by mere routine, without knowing the foundation or reasons, cannot deviate from the narrow path in which he is confined, when any unexpected accident occurs.

In the following treatise, we have endeavored to explain these principles, as far as they have been connected with the subjects upon which it treats, and to illustrate them in such a manner as to be easily understood by the general reader.

The first part of the work we have devoted to the construction of Conservatories, Graperies, Green-houses, Pits, Frames, and every kind of horticultural buildings, giving the different positions and aspects most suitable to each, and the various purposes for which particular structures are best adapted. We have also fully considered the different kinds of materials generally used in the erection of these buildings, and the respective merits of each. Glass, and its influences on vegetation, are also fully considered and discussed in this part, — a subject which has hitherto received very little attention from horticultural writers, but is nevertheless one of the most important items connected with exotic gardening. We have given the useful experiments of Mr. Hunt, on LIGHT, and its effects on vegetation and germination, and all other information which we have considered useful on this part of our subject.

The second part embraces the most approved methods of heating horticultural structures, giving the principles of combustion and consumption of fuel, the prevention of smoke, and the various volatile products of the coal ; the construction of flues and furnaces ; the different sizes and heating powers of pipes and boilers ; the circulation of water, and the peculiar modifications of apparatus suitable for particular structures. We have given a considerable number of illustrations in this part, showing various methods of heating, with all of which we have had extensive practice, and some of them on entirely new principles. The various merits of hot air and hot water are considered on scientific as well as on practical grounds, and each acknowledged for what it is worth.

The third part may be called the theory and practice of

2

ventilation, including some valuable investigations of the physiological effects of the atmosphere, under different circumstances,
and at different temperatures. Many of our remarks nave
assumed a greater length than we originally intended, and if
some appear repetitionary, this is in order to avoid, as much
as possible, all strictly scientific technicalities and abstruse
reasoning, whereby the minds of practical men are frequently
unable to understand fully the end to which you direct them.
We have added a section on the protection of horticultural
structures in severe weather, — a subject which is worthy of
much consideration.

I may observe, that, in pointing out and freely commenting on principles and practices which are erroneous, but which
have been practised and promulgated by others, it is under the
impression that such errors, carrying with them, in general,
some plausibility, have led, and may still lead, others to fall into
similar mistakes. However invidious, therefore, be the task of
pointing out these errors, it would be manifestly impossible to
write on this subject without noticing them, and, if possible,
pointing out the difference between right and wrong. This is the
only apology which can be offered for the freedom with which
some of the opinions and methods of others have been com-
mented on in the various parts of this treatise. We have, however, expatiated on them candidly, and in the true spirit of
inquiry, pointing out the applicability of their principles and the
utility of their practice.

The different parts of the subject have been arranged under
different heads, as far as has been practicable, in order that any
of the different parts may be pursued intelligibly and clearly.
In extenuation of any errors which may be found, we hope it
will be considered that many of the points treated on are
entirely new, and as yet undeveloped ; that no comprehensive
view of the principles of exotic culture has yet been given.
But we must not be understood to offer excuses for any errors
other than those that are embraced by this extenuating clause
which will be acknowledged if rectified in the true spirit of
philosophical inquiry.

PART I. CONSTRUCTION OF HOT-HOUSES.

SECTION I.

SITUATION.

1. *Site and position.* — Before proceeding to details regarding the structures themselves, it will be necessary to consider, briefly, the situation on which the structures are to stand. A glazed structure depends for its effect very much upon its position; and as the position 'most desirable for effect may very possibly militate against the utility and efficiency of the structure, the question presents a double claim to our consideration. In illustrating the position most desirable for the erection of houses for horticultural purposes, I assume that the paramount object is utility. I will subsequently point out reasons which frequently occur to render the position of green-houses and conservatories beyond the control of the erector.

By site and position I must not be understood to imply merely the aspect upon which a house for horticultural purposes should stand. The aspect of a house may be affected by circumstances which have no relation to its site. In other words, the glazed elevations of a house may be turned in any direction, while the position may be altogether unsuitable whichever aspect may be given to it. The weather, at all seasons of the year, has undeniably more influence on a house in some situations than it has upon houses in others more favorably placed; and this influence is sensibly felt by the products which are grown within them.

The climate, and especially the prevailing winds of the locality should be studied attentively, in order to anticipate their changes, and avoid, as far as possible, their injurious effects. No doubt it is sometimes difficult to ascertain the precise spot on which to erect hot-houses, with these considerations in view, particularly when the ground is extensive and the choice limited; yet, in most places, there are some spots preferable to others. A bleak, elevated position should never be chosen, if there be any choice left. If a bare, elevated spot *must* be chosen, either on account of there being no alternative, or from other adventitious considerations, such as to obtain a commanding view of the surrounding country, or to present a more imposing appearance from the mansion, or from any other point of sight from which it may be thought desirable to view them, then the background should always be planted up with trees. This is indispensable, for two important reasons : —

(1.) *For shelter.* The northern winds are cold and biting in frosty weather, and air can be admitted when the houses are well sheltered, when it otherwise would be impossible to do so without injury to the plants. Moreover, the north side of a horticultural structure of any kind is the only one that can be appropriately sheltered with tall growing trees. It is, therefore, the more necessary that trees should be planted close enough to break the wind, but not so close that their overhanging branches, when they have attained their full size, may drip upon the glass. This last is an evil which ought, in all cases, to be avoided. Neither ought new houses to be placed so near trees, already standing on the grounds, that these circumstances may occur.

(2.) *For beauty and effect.* I do not mean, in this paragraph, to allude to hot-houses in general as handsome architectural objects in the grounds of a country residence, — to which consideration I will subsequently allude, — but merely to the effect which hot-houses of the cheapest and plainest description may be easily made to produce, without much trouble or expense, or without adding one cent to the cost of the structure itself. Let any person take a glance at a structure of glass, or range of such structures, having nothing but the distant sky for a background, and compare it with another, resting upon the green, glossy foliage of luxuriant trees towering above them, and these again reflecting their irregular outlines against the cloudless horizon behind them, and' he cannot fail to be struck with the tame and spiritless appearance of the former, and equally, also, by the picturesque and pleasing effect produced by the latter.

A conservatory, or green-house, avowedly ornamental, and intended as an object of architectural beauty, or of individual elegance, requires the most exquisite taste and skill in harmonizing the objects around it. These surrounding objects, whether for utility or embellishment, may be so arranged as to heighten the effect of the whole, without impairing the individual effect of the structure, or hiding any of its beauties. The various features of the structure should be presented to view from different points; and if, from any walk or portion of the grounds, the structure present rather an unfavorable aspect, then some object should be interposed to obstruct the view from this particular point. When a walk is led along the skirt of a wood or plantation, where a glimmering of the structure is continuously visible from among the trees, the effect is bad, and ought, by all means, to be obviated by planting shrubbery and underwood, leaving here and there an open vista through which a full view of the whole building, or portion of it, may be obtained.

It has, for some time, been the rage in this country to place horticultural buildings of all kinds upon eminences, and surround them, either wholly or in front, with square terraces. These terraces are made sometimes of brick, in all its primitive redness, sometimes of small stones and mortar, and more frequently, perhaps, of grass, nearly perpendicular. It is generally difficult to discover which is the most unnatural and unsightly; and, in nineteen cases out of twenty, we have found the terrace itself, of whatever materials, of very questionable taste. Terraces grew out of necessity, — not out of taste, — except, perhaps, in the Dutch school, which an able writer on this subject styles "a double-distilled compound of labored symmetry, regularity, and stiffness." * A terrace may be in very good taste, in connection with a pretty little Tuscan or Italian

* Downing's Landscape Gardening.

2*

villa, when it is finished and ornamented as a terrace should
be, *i. e.*, with vases, urns, &c., of sizes and forms harmonizing
properly with the architecture of the building. The same principle may be applied to detached conservatories when placed in
the grounds as ornamental objects.

While speaking of terraces, it may not be out of place to
remark, that, about some of the finest gardens of this country,
these grass walls are introduced to absolute satiety. Nothing
like a gentle, undulating surface is for a moment tolerated, but,
as a matter of custom, the ground must be levelled, and flanked
by a terrace. Now we think that when terraces are found necessary in front of a garden structure, of an ornamental character, they ought to be of a different character from those intermi-
nable sod banks so liberally constructed about some fine places
that we could mention, but forbear doing so, on the principle,
that, where much has been done, a few errors in taste may be
justified. However, it cannot be denied that a steep bank of
grass, twelve or twenty feet deep and as many from the walls
of the building, void of any architectural decoration or ornament
of any description, save its own unrelieved formality, is in as
bad taste as would be the surrounding of a mud-walled hut with
architectural balustrades and sculptured ornaments. Steep,
formal terraces, without architectural decorations to unite and
harmonize them with the structure, are, unquestionably, the
most insipid and meaningless objects that can be introduced
into ornamental grounds.

What is called an architectural terrace, consisting of a low
parapet and balustrade of handsome masonry, or other rich
ornamental work, has always a pleasing effect, especially when
attached to buildings of an ornamental character,* whether
these buildings be for dwellings, or for horticultural purposes.
These terraces, however, are very different from those perpendicular turf-banks, of which I have already spoken. The
former are truly artistical, and, in connection with classical

* The reader who is interested in this subject, and wishes for further
information on this kind of ornamental terraces, is referred to the elegant remarks and illustrations thereon by Mr. Downing. [See *Downing's
Landscape Gardening ; section, Architectural Embellishments.*]

structures, constitute the harmonizing link between art in the building, and nature in the grounds. The latter are neither artistical nor natural.

Let the reader fancy to himself a horticultural structure, of unusually large dimensions, situated on the southern declivity of an open field, without a single green leaf of foliage to intervene between the unbroken whiteness of the structure and the distant sky. The very ornaments of the building are altogether hidden, even at the distance of a dozen yards, because their forms are viewed upon a background of cloudless vacancy; directly in front is a terrace, more than a dozen feet deep, and so steep as to require a ladder to scale it, and at the bottom it terminates with an abrupt angle, adjoining a potato and cabbage garden. However unquestionable may be the position of the splendid structure here referred to, there is something so irreconcilably incongruous about its precincts, that the most untutored imagination is at once struck with the total want of harmony, unity, and effect. The terrace itself has the unfinished appearance of a dwelling-house, where the work has been suspended before the roof and chimneys had been put on; a thing appearing to have an isolated and independent existence, having no apparent relation either to the structure or the grounds, and heartily despised by both. Now the *position* of the building referred to is, undoubtedly, excellent, and a better site could not be found, to produce a more imposing effect from a front view, which, in horticultural structures, is generally the best, providing the structure be sufficiently elevated above the axis of vision. But in the present case the effect is destroyed; first, by a total want of unity and harmony in the foreground, and, secondly, by a want of the deep, dark foliage of trees, presenting their irregular outlines against the sky in the background, which gives all buildings, and more especially those of a light character, as hothouses, &c., that picturesque and pleasing appearance, particularly when the surface of the ground is broken by undulations, and the scenery diversified with a variety of objects, distinct in themselves, yet harmonizing with each other.

From the foregoing observations, the propriety will be seen of placing horticultural erections in the immediate vicinity of

large trees, and of raising them where they do not already exist.
A beautiful writer on this subject has observed, that green-
houses in the country, without trees about them, are like ships
divested of their masts and rigging, and impress the mind with
the idea of their having wandered from their right position;
and, as Loudon justly remarks, a tree is the noblest object of
inanimate nature, combining every species of beauty, from its
sublime effect as a whole, to the most minute and refined expression of the mind.*   We cannot too strongly urge the propriety of choosing a site where these advantages may be gained.
This branch of landscape gardening has been already treated in
a masterly manner by various writers; therefore we consider it
unnecessary to dwell any longer upon it. †

The choice of position may, in some instances, be decided by
other circumstances, such as an abundant supply of water. This
is indispensable, in hot-houses of every description, though it
seldom forms a very important consideration with architects, in
their designs, who are perfectly unconscious of the amount of
labor and expense subsequently created by a deficiency of this
element.   It is, therefore, desirable that the site chosen should
command a plentiful supply of water, at all seasons of the year,
independent of what may be collected from the roof.   It should
be considered that the period when the largest quantity of water
is required for the use of the plants, is also the time when the
supply from rains is scantiest and most precarious; and though
ample provision must be made for collecting all the water that
falls upon the roof, into tanks and reservoirs, suitably and conveniently placed for that purpose, yet this supply is not to be
entirely relied upon; and hence water ought to be conveyed by
pipes, or some other means, from the nearest source, to supply
the tanks when the rain-water is exhausted.

Where a stream of water is commanded by the position of

---

* Loudon's Encyclopedia of Gardening.

† Those who wish to study the principles of landscape gardening, will
find all that is requisite for their instruction and improvement in "Downing's Landscape Gardening," the only work we know wherein the principles of the art are treated in such a manner as to render them perfectly
applicable to this country.

the structure, it would be most desirable to convey it through the interior of the house, in a kind of rill, or small stream, running through a shallow channel, or, what would be still better. to fall into a tank, over a small precipice, forming a little cascade, or water-fall. If the stream had sufficient power by its declivity, a small jet might be kept continually playing. In an ornamental plant structure, this would be the *ne plus ultra* of a water supply; besides, the house would be kept delightfully cool in the hottest days of summer, and the rippling of the stream over the cascade, or the playing of the fountain, would prove the most agreeable music to the ear in the hot days of summer.

We have here alluded to water, merely in so far as it may be likely to affect the choice of position. Of course, water may be supplied to a house by various other means, such as force pumps, and that admirable invention, the water-ram, by which jets, cascades, &c., may be also obtained; but all these are attended with considerable expense, as well as subsequent labor, and, therefore, a natural, constant, and abundant supply of water, when possible, should not be abandoned, even at the expense of some trifling advantages in other respects. We have known places where the labor of carrying the water for the different departments of the exotic establishment during summer, exceeded the labor required to keep the garden in order.*

In regard to the precise elevation best suited for the site of horticultural buildings, various opinions exist; some prefer low-lying grounds, others prefer a considerable altitude; we have frequently seen both parties run into extremes. Low situations are generally warmer, and better sheltered from boisterous winds, which, however, is more than counterbalanced by certain evils consequent upon a very low site. In spring, low, swampy places are always subject to heavy depositions of dew and mist, which render them cold and damp, and expose vegetation of every description to be destroyed by vernal frosts, which is avoided in more elevated situations. We have this spring had abundant evidence of this fact, in a very large tree of the Platanus Occi-

* For further information regarding cisterns and supplies of water, see Sec. IV., Internal Arrangements.

dentalis, which had its leaves entirely destroyed, after they were fully expanded, and are now strewed upon the grass beneath. This tree, with various others that shared the same fate, stood in a low part of the pleasure ground, beside a lake. Trees of the same species, on higher ground, escaped without injury.

We have invariably found, in our experience, that plant-houses situated in very low grounds, were cold and damp in winter, and hard upon the more tender kinds of plants. In summer, the atmosphere is generally stagnant and unhealthy, to plants as well as animals. If circumstances, therefore, afford any choice, very low situations should be avoided, as it is more easy, and certainly more profitable, to bring an elevated and airy situation into the condition desired, than it is to obviate the injurious effects of a low one.

2. *Aspect.* — We find that most people prefer a southern aspect for their hot-houses, *i. e.*, placing the front elevation due south. The absolute propriety of this preference, however, deserves to be questioned, as experience has taught us that some valuable advantages are gained by placing hot-houses, for the growth of fruits, on a south-eastern aspect. Let it be observed, that we are alluding at present to what is termed lean-to, or shed-roofed houses, *i. e.*, houses having only one sloping side, — a kind of structure still generally used for the production of grapes, &c., during the early part of spring, and which are probably better adapted for that purpose than span-roofed houses. In fact, we should prefer a south-eastern aspect for lean-to houses, whether they were intended to grow fruits or flowering plants; for, even in this clear and comparatively cloudless climate, this aspect has advantages which, in our opinion, are not possessed by any other; and, indeed, the greater intensity of the sun's rays at midday here than in England, gives this aspect greater advantages in this country than in any other where the sun is less powerful. The morning sun is more strengthening and exhilarating to plants than during any other period of the day, and more especially to plants kept in houses without artificial heat; but the same argument holds good in all houses. We find that

hot-houses, even during the early part of summer, — except fire heat be maintained, — sometimes fall exceedingly low at night, and become cold and chilly, with the aqueous vapors contained in the atmosphere, by the high temperature of the preceding day, condensed into water by the low temperature of the night, and depending in small globules from the leaves of the plants, the under surface of the glass, and other parts of the house, rendering the approach of the sun's cheering beams, a few hours earlier than if the house were placed meridionally, above all things acceptable.

It might be plausibly argued, that, if we take the south-eastern aspect for the purpose of gaining the morning sun, we must lose it for the same period in the afternoon, which, altogether, makes it the same thing to the house. This is not true in practice ; though the period of the sun's duration upon the house in both cases be the same, yet the advantage gained, by taking the morning and losing the afternoon sun, is very great. The rays thus lost in the evening are of little consequence compared with those gained in the morning, because the plants are then partially enfeebled, and their elaborative powers impaired, if not altogether suspended, by the strong midday heat. By various experiments on the shoots of young plants, we found that their elongation was greatest during the mild hours of the morning, before the sun had attained its meridian fierceness.

In general, we find that plants are more prostrated by the influence of the afternoon sun than during any other period of the day, and it is supposed, by many, that the sun's heat is more powerful and oppressive in the afternoon — that is to say, from one to three — than it is when on its meridian: However this fact may be scientifically supported, it certainly holds good in experience.* Supposing, then, that such is the case, we con-

* This may be accounted for by the air having been already warmed to a high temperature, by the sun acting upon it during the previous part of the day ; and, the deposited moisture of the preceding night having been already evaporated from the surface of the earth, the lower strata are highly rarefied. The hot sun, continuing to act upon the lower stratum of air and the dry surface of earth, gives the former that languid, oppressive, and suffocating character, which is experienced by every one.

sider it another fact in favor of a south-eastern aspect, as the sun's rays will thus be made to strike the roof more obliquely, and will be less likely to scorch, or otherwise injure, the plants, than if shining perpendicularly to the plane of the roof.

Many authors might be quoted, in support of a south-eastern aspect; and one of the best garden authors, of his own or any other time, says, "An open aspect to the east is a point of capital importance, on account of the early sun." When the sun can reach the garden at its rising, continuing a regular and gradual influence, increasing as the day advances, it has a gradual and most beneficial effect in dissolving the hoar frost that may have been deposited the previous night. On the contrary, when the sun is excluded till about ten in the morning, and then suddenly darts upon it with all the force derived from its increased elevation, and increased power, it is very injurious, especially to fruit-bearing plants, in the spring months. The powerful rays of heat at once melt the icy particles, and, immediately acting upon the moisture thus created, scald the tender blossoms and leaves, which droop and fade as if nipped by a malignant blight.*

These remarks, it is true, are by an English author, and have reference to the climate of England; but they apply to us in full force in this country, and, in many locations here, are still more applicable than to any country in Europe.

The morning sun is not only more agreeable to vegetable as well as animal development, but, as we have already observed, vegetation proceeds more rapidly under its influence than it does during any other period of the day. This may be accounted for by the fact, that the nourishing gases have been accumulating during the partial suspension of elaboration in the night, and, on the approach of the sun's vivifying beams, these functions are resumed with increased activity, and continue so, under the mild influence of its less powerful and fierce effulgence, until their energies are paralyzed by its burning rays, at midday, when they make little more progress till the next morning.

We have heard similar arguments adduced in favor of a south-

---

* Abercrombie's Practical Gardener.

western aspect for late houses, and these facts have regulated the erection of some extensive houses with which we are acquainted. In this country, however, hot-houses are seldom erected for the express purpose of retarding grapes, or other fruits, although we have no doubt that very late grapes would pay better than early ones, since there would be very little expense in their production. The sun's rays in this climate are so powerful, that the difference in aspect may not be so perceptible, in regard to late and early forcing, as in England; still we have no doubt the difference will be found sufficient to justify the erection of houses for these purposes, on the aspects we have pointed out as being most suitable for each.

In the erection of span-roofed houses, that is, houses with double roofs, it makes very little difference, in the opinion of many, which way the house may stand, and, upon the whole the arguments hitherto used, in favor of one aspect over another, have been so feeble as hardly to deserve any consideration. Supposing the house to be a parallelogram, or long square, with both gables glazed, as well as the sides and roof, then, we think, it may stand any way in which the nature of the site, or taste of the erector, may dictate. Light being the most important point of attention in the construction of hot-houses, these are better adapted for plant-growing than those whose transparent surface forms only a segment of their transverse section.

As a general principle, provided other circumstances are favorable, we would recommend the house to stand north and south, with its longer elevations towards the east and west; we find this to be the opinion of some of the best gardeners in the country, with which we fully agree. If any advantage be gained by placing the house in one direction, in preference to another, we think it is the one mentioned, as the rays of the meridian sun will then strike the glass in an oblique direction, and have less power than if they were to fall upon the glass at right angles to it.*

The aspect of conservatories attached to dwelling-houses

* For more detailed information on this matter, see Sec. II., Design and Slope of Roof.

3

must be regulated by the position of the building, or the fancy
of the architect.   These are deplorable erections, generally;
nine tenths of them unsuitable, in the superlative degree, not
for want of cost, but for want of skill.   As the remarks we have
to make, on this part of our subject, belong to the next section, we will just add here, that a conservatory ought never
to be placed on the northern aspect of a building, nor situated
in such a manner, in relation to the dwelling-house, that the
sun's rays may be prevented from falling on the conservatory
during at least one half the day.

SECTION II.

1. *General Principles.* — To ascertain principles of action, it is always necessary to begin by considering the end in view. The object or end of hot-houses is to form habitations for vegetables, and either for such exotic plants as will not grow in the open air of the country where the structure is to be erected, or for such indigenous or acclimated plants as it is desired to force or excite into a state of vegetation, or accelerate in their progress to maturity, at extraordinary seasons. The former class of structures are generally denominated green-houses, or botanic stoves, in which the object is to imitate the native clime and soil of the plants cultivated; the latter, comprehending forcing-houses and culinary stoves, in which the object is to form an exciting climate and soil on general principles, and to imitate particular climates.

The chief agents of vegetable growth in their natural habitations are *light, heat, air, soil,* and *moisture;* and the merit of managing these structures, and the success of cultivating vegetables in them, depend on the perfection with which nature in these respects is imitated.

To carry out the imitation to perfection, or anything like an approach to it, it is absolutely necessary, as we have previously observed, to be acquainted with the nature and habits of the plants under cultivation. Vegetable physiology ought to form a part of the acquirements of the hot-house architect; and the chief cause of the great improvement in these structures, of late years, in England, is traceable to the fact, that their erection is no longer left, as formerly, under the control of mansion architects, as they are at the present day throughout the length and breadth of the United States; and the chief reason why we see horticultural structures erected so numerously in this country, in violation of the first principles of plant culture, is undoubtedly due to the same cause. The conservatory is generally left to the uncontrolled management of the architect, who, of course, makes this structure to correspond with the rest of the building, without giving the slightest consideration to the vegetable beings that are to grow in it. If we consider this matter in its different bearings, giving to professional architects the justice which is due them, it would be somewhat unreasonable to expect them to plan conservatories otherwise. An architect is, by education, taught to study and apply principles in his art, which, when carried into effect, as we sometimes see them in the construction of plant-houses, are in direct opposition to those laws which nature has laid down and determined as essential to the vigorous development of vegetable life. Can it be expected, then, that an architect will tamely surrender the grand principles of his art, — the antiquity of which is coeval with Cheops, and which has been the boast and pride of the greatest empires of the old world, — in meek submission before the yet half-developed principles of vegetable physiology, or even to the humble dictates of practical gardening? To expect such a concession would be tantamount to expecting an architect to build dwelling-houses with drawing-rooms solely adapted for the accommodation of plants, altogether irrespective of other purposes to which drawing-rooms are generally applied. Hence, we find the conservatory placed just where it is most subservient to the general design of the mansion, most frequently in a corner or recess of the main building, having two or three sides of solid opaque material! To civil architecture, as far as respects mechanical principles, or the laws of the strength and durability of materials, they are certainly subject, in common with every other species or description of edifice; but in respect to the principles of design and beauty, the foundation of which we consider, in works of utility at least, to be "fitness for the end in view," they are no more subject to the rules of civil architecture than is a ship or a fortress; for those forms and combinations of forms, and that composition of building, which is very fitting, and, perhaps, beautiful, in a habitation for man or for domestic animals, is by no means fitting, and consequently not beautiful, in a habitation for plants. Such, however, is the force of habit and professional bias, that it is not easy to convince architects of this truth. Structures for plants are considered by them no further beautiful than as they display something of architectural forms, which, according to the innumerable illustrations presented to us, consist of a solid opaque building; for it is an undeniable fact that what are called fine architectural conservatories, are designed, not for the purpose of growing or exhibiting flowering plants, for they have not even the appearance of adaptation for this purpose. One half of their entire surface is obscured by pilasters, blocking-courses, cornices, projections, massive astragals, sash-bars, etc., until the transparent glass forms only a small fraction of the surface ostensibly appropriated to the transmission of light. To complete the opacity of the structure, the whole is obscured or shaded one half the day by the main building. There is no ideal exaggeration here; they form the grand rule in ornamental conservatories, and the exceptions are few. Let us take, for example, the splendid conservatory erected by J. W. Perry, Esq., at his mansion at Brooklyn, near New York, and figured in Downing's Landscape Gardening, which is extolled as one of the most beautiful conservatories in the country, and with some degree of justice, for it is both more beautiful and better adapted for the purpose than many others to which we can allude. Yet, beautiful and fit as it may be considered, there never was, and never will be, a plant grown in it to perfection, nor is it possible by any species of care or skill to do so in such a structure.

We have taken the liberty of particularizing this conservatory, because it has been made the model of various others which we are acquainted with; and we justify our allusion to it on the following grounds: because the house in question has been figured and commended by such an able authority, and consequently been regarded as a model of perfection by many who know not the difference between a structure "fit for the purpose," and one merely beautiful in itself; and, moreover, because we are well acquainted with the structure itself, as well

3*

as with the able and excellent gardener who has managed it for
many years, and who finds it impossible to grow plants within,
and has long since given up the case as utterly hopeless; the
only result which could be expected.

It may seem strange that ten or twelve thousand dollars
should be expended upon a plant-house, and, after all the
expense, the house be unfit for the growth of plants, and that
this fitness could be more extensively obtained at one twentieth
the cost.   Such, however, is the case, and will continue to be
so, till the design be considered in relation to "fitness for the
end in view;" and that this is far from being the case, we have
lately experienced sufficient proof.   Buildings like that we have
just alluded to, may properly be called beautiful specimens of
architecture, but if the principles of design or beauty be regarded
on fitness for the end in view, — as we believe it to be in works
of utility, — then, as plant-conservatories, these structures ought
to be condemned.

I have no doubt some of our architectural readers, and lovers
of dull, massive, gorgeous, and grotesque conservatories, will
pronounce against such a violation of the principles of architecture, as would undoubtedly be perpetrated by building a mere
shell of glass to form a counterpart of the solid masonry of a
large mansion.   Conversing on this point lately with a talented
architect, he said, " Conservatories can never be reconciled with
mansion architecture if they must be erected upon such principles; the thing is utterly inconsistent with beauty in a building.
Such an appendage," said he, " would be as absurd as putting a
gauze covering over a buffalo robe to withstand a snow storm."
It would be useless here to reply to the injustice and inapplica-
bility of these observations, and we will let them go for what
they are worth.   They serve, however, to convey a pretty accurate idea of the estimation in which architects hold the principles
of plant culture, even when pointed out to them; or, as we might
term it, how little they care for the beauty expressed by "fitness
of purpose."   Utility, however, is undoubtedly the basis of all
beauty in works of use, and, therefore, the taste of architects, so
applied, may safely be pronounced as radically wrong.

2. *Light.* — In erecting horticultural structures of any description, the first and decidedly the most important object to be kept in view, is the introduction of light; and really, though this point presents itself to architects in its simplest and plainest reality, it appears to be scarcely ever fully considered; at least, we are induced to conclude so, from the instances already before us. It is easy for any person to satisfy himself of the wonderful effects of light upon vegetables under artificial culture, by the most familiar illustrations. When plants are placed against a wall, or other opaque body, they will speedily turn the surface of their leaves to the light, although the medium of its entrance should be many yards distant. One of the principal reasons why plants thrive so badly in dwelling-houses, is in consequence of their being deprived of that supply of light which is essential to their development. Set a plant how or where you will, it will twist and turn itself in any direction for the purpose of presenting its leaves to the light, or to the aperture where it enters unobstructed. Pure air is also a most essential element in the economy of vegetation; but we may safely assert, after much experience, that plants under artificial culture suffer far more from a deficiency of light than from a deficiency of what is called pure air. The reason of this appears obvious. By the latter deficiency a plant is merely deprived of its necessary food; but by the former deficiency the plant is entirely deprived of its vegetable functions, or its energies are so enfeebled as to be incapable of assimilation. We are not speaking here of light merely as distinguished from darkness, for we are told, upon good authority, that the luminiferous ether is radiated in all directions from its grand source, viz., the sun,* but of its properties and influence on plants when transmitted through a transparent medium, such as glass. Every gardener knows that plants will not only fail to thrive without much light, but will not thrive unless they receive its direct influence by being placed near the glass. The cause of this last fact has never been satisfactorily explained. It seems probable that the glass,

* Principles of Chemistry, by Prof. Silliman, Jr.

acting in some degree like the triangular prism, **partially decom-**
poses or deranges the order of the rays.

The theory of the transmission of light through transparent
bodies is derived from the well-known law in optics, that the
influence of the sun's rays on any surface, both in respect to
light and heat, is directly as the sine of the sun's altitude,
or, in other words, directly as its perpendicularity to that surface. If the surface is transparent, the number of rays which
pass through the substance is governed by the same laws.
Thus, if one thousand rays fall perpendicularly upon a surface
of the best crown glass, the whole will pass through, excepting
about a fortieth part, which the impurities of even the finest
crystal, according to Bouquer, will exclude. But if these rays
fall at an incidental angle of 75°, two hundred and ninety-nine
rays, according to the same author, will be reflected. The incidental angle, it will be recollected, is that contained between the
plane of the falling or impinging ray and a perpendicular to
the surface on which it falls.*

In building a green-house or conservatory, then, light ought
to form the first point of importance, as success in plant culture
is entirely subservient to it, and we know full well, from experience, that no skill, however perfect, and no attention, however
zealous, will compensate for a deficiency of light. Indeed, no
contingent or permanent advantage can justify, to the mind of
the experienced gardener, the adoption of one inch of opaque
material in the sides and roof of a horticultural building; and
no part of the structure, from the side-shelves and upwards,
should be rendered opaque that can consistently be covered with
a material capable of admitting the rays of light. For pillars
and other appendages of strength, the material ought to be as
light as is consistent with strength and durability in the structure; and, although we do not recommend such an erection
adjoining a dwelling-house, experience has taught that, both in
this country and in others, a mere shell of glass, so to speak, is
not only the cheapest, but also the best adapted for the artificial
culture of a l kinds of plants, both for fruiting and flowering;

* See Inclination of Hot-house Roofs.

i. e., plants that are cultivated solely either for their flowers or for their fruit.

The exact manner in which light acts upon plants has been studied by Dr. Daubeny, and others, and especially Mr. Hunt. The result of these inquiries is given thus, in the Gardener's Chronicle, of August 16th, 1845 : "Assuming, with Sir David Brewster, that the prismatic spectrum consists of only three primitive colors, namely, red, yellow, and blue, it is ascertained, by experiment, that the maximum of *heating power* is found on the confines of the *red rays ;* that the largest amount of light is given by the *yellow rays ;* and that the chemical power exists most strongly amidst the *blue rays*, of the spectrum. If we take a deep red glass, which has been colored with the oxide of gold, it will be found that the quantity of light which passes through it is very small ; and, by using photographic paper, it may be ascertained that the amount of that principle which produces chemical change is also very little, whereas the heat rays suffer no interruption. A deep yellow glass, or a cell filled to the thickness of an inch with a solution of bicromate of potash, intercepts the chemical rays, but admits of the permeation of all the luminous rays, and offers but little interruption to the calorific rays. If, however, we cover a pane of this yellow glass with another of pale green bottle glass, the passage of the heating rays is much impeded. A deep blue glass, such as is used for fingerglasses, colored with oxide of cobalt, or a solution of oxide of copper in ammonia, has the property of admitting freely the passage of all the chemical rays, whilst it obstructs both the heat and light radiations. Experiments conducted with colors thus obtained led Mr. Hunt to the following conclusions :

(1) *Light which has permeated* YELLOW *media.* LIGHT RAYS.— In nearly all the cases the germination of seeds was prevented, and even in the few cases where the germination was commenced, the young plant soon perished. The germination seemed referable to the action of the heat rays which had passed the medium employed, rather than to the light. Agarics, and several varieties of fungi, flourished luxuriantly under this influence. Although the luminous rays may be regarded as injuri-

ous to the early stages of vegetation, Mr. Hunt believes that,
in the more advanced periods of growth, they become essential
to the formation of woody fibre.

(2.) *Light which has permeated* RED *media.* HEAT RAYS. —
Germination, — if the seeds are very carefully watered, and a
sufficient quantity of water is added to supply the deficiency of
the increased evaporation, — will take place here.   The plant is
not, however, of a healthy character, and, generally speaking,
the leaves are partially blanched, showing that the production
of chlorophyl is prevented.   Most plants, instead of bending
towards red light, as they do towards white light, bend from it
in a very remarkable manner.   Plants, in a flowering condition,
may be preserved for a much longer time, under the influence
of red, than under any other media ; and Mr. Hunt thinks
that red media are highly beneficial under the fruiting process.

(3.) *Light which has permeated* BLUE *media.* CHEMICAL RAYS.
— The rays thus separated from the heat and light rays, and
which Mr. Hunt has proposed to call ACTINIC, have the power
of accelerating, in a remarkable manner, the germination of seeds,
and the growth of the young plant. After a certain period, varying with nearly every plant upon which experiments have been
made, these rays become too stimulating, and growth proceeds
rapidly, without the necessary strength.   The removal of the
plant into yellow rays, or into light which has penetrated an
emerald green glass, accelerates the deposition of carbon, and
the consequent formation of woody fibre.   It was also found
that, under the concentrated actimic force, seeds will germinate
beneath the soil, at a depth in which they would not have grown
under natural conditions.   Mr. Hunt believes that the germination of seeds in the spring, the flowering of plants in summer,
and the ripening of fruits in autumn, are dependent upon the
variations in the amount of actimism, or chemical influence, of
light and heat in the solar beam at these seasons.

It must, however, be observed, that, although such experiments
have much physiological interest, the value of them is greatly
diminished by the necessarily imperfect manner in which the
prismatic colors are separated by artificial preparations.   It is

almost, if not quite, impossible to form *pure* colors artificially. The yellow, for instance, of the bichromate of potass contains both red and violet in abundance.*

It has been already ascertained that the amount of assimilation, and consequently of the healthy exercise of its vital functions, depend upon the intensity of the light to which the plant is exposed. In bright sunshine they perspire most; in weak diffused light, and in darkness, none at all. Hales found that a cabbage lost nineteen ounces of weight per diem, and a sunflower twenty. He estimated the average rate of perspiration by plants to be equal to seventeen times that of a man. In one of his experiments he found that the branch of an apple-tree, two feet long, with twenty apples, exposed to bright sunshine, raised a column of mercury twelve inches in seven minutes. But a dry, arid atmosphere, especially if in motion, also robs the plants of their moisture independently of light.

The clear and unclouded skies of this country do not, as some suppose, obviate the necessity of surrounding the plant with a transparent medium in all directions, nor does the *dark and sunless* climate of England render it necessary that the houses should be more transparent there than here. It is a practical absurdity to fancy that in England there is less *light* than in this country, and that, because the mid-day sun is more powerful, they can do with a greater opacity of structure. Those who make such statements manifestly know as little of the climate of England as of the natures of its skies, and mislead those who know as little as themselves. No argument whatever, based upon the brightness of the sunshine at mid-day, can serve to justify the adoption of one single inch of opaque material in a horticultural building. It is very easy to reduce the quantity of light, or break the rays of the sunshine, by shading; but it is not so easy to increase the quantity of light in the dark and gloomy months of winter; and such sort of plant-houses will damp the energies and zeal of the most skilful gardener, as well as his tender exotics. When he sees these errors, which he cannot remedy, and observes his plants speaking in a language which

---

* For further experiments on Light, see Sect. IV., *Glass*.

cannot be mistaken, even by the most inattentive, " Give us light, or we shall die ! " he gives up their case, in hopeless despair, as being altogether beyond his control. And thus we have known excellent gardeners censured for neglecting things, and for doing badly what it was not in their power to do better.

Solar influence being necessarily connected with the roofs of hot-houses, we will discuss these subjects in their relation to each other, including inclination and reflection, in the following sub-section.

3. *Slope of hot-house roofs.* — In regard to the theory of the transmission of light through transparent bodies, we have already stated that the influence of the sun's rays on any surface is directly as his perpendicularity is to that surface ; and, according to Bouquer, that if one thousand rays fall perpendicularly upon a surface of glass, the whole pass through, excepting about twenty-five rays, or one fortieth part of the whole. But falling on the same surface at an incidental angle of about 75°, then two hundred and ninety-nine, or nearly one third of them, will be reflected. The influence of the sun on the roofs of hot-houses depends very much on the principle there given, — at least, so far as regards the form of its surface. This principle has been applied, in various ways, for the purpose of obtaining the full influence of the sun's rays at certain seasons of the year. We have managed forcing-houses where the roof was laid at right angles to the sun's rays in mid-winter, — the period when the most powerful rays were required for forcing purposes.

Although it cannot be denied that much more depends on the management of the house, for the success of cultivation, than on the inclination of the roof, yet it is the most satisfactory method to proceed on what may be considered something like principles. And in this country we find this the more necessary, because the heat of the sun's rays, at certain seasons of the year, is so violent as to prove injurious to vegetation under any circumstances. And hence, this principle should be adopted in the construction of hot-house roofs, that their perpendicularity to the sun's rays, at the hottest period of the year, should by all means be avoided.

In England, the most common elevation of roof is an angle of 45°, which, in the latitude of London, would form a perpendicular to the impinging ray, about the beginning of April, and the beginning of September, — which also makes the obliquity of the rays greatest when they are most powerful, viz., during the month of June. "This angle is preferred by most gardeners," observes Loudon, "probably from habit." We think, however, that something more than mere habit justifies the adoption of this angle, — more especially for forcing-houses, — since by it the benefit of perpendicularity is obtained at a period when the rays are comparatively feeble and most necessary.

Fig. 1.

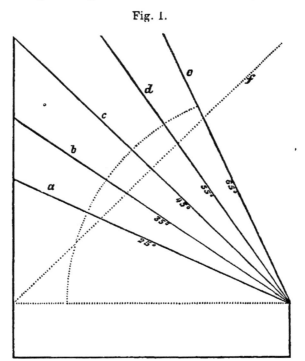

As some of our readers may not have made themselves sufficiently acquainted with the altitude of the sun in relation to the slope of hot-house roofs, we have annexed the above figure, (Fig. 1,) which represents the slope of five different roofs on the angles marked by their respective complements. *f* represents

4

the altitude of the sun in the latitude of London, the impinging ray falling on the roof, c, at an angle of 45°. It will be seen that the angle, contained between the back wall of the house and the inclined plane of the roof, c, is just equal to the sun's altitude, — the one forming an exact perpendicular to the other.

Allowing, then, for the difference of altitude betwixt the latitudes of London and Philadelphia, for instance, we have a difference of inclination of about 11°. Hence the roof of a hothouse, to receive the same influence of the sun's rays at that period, would be at an angle of 34°. The difference will be more closely perceived by the following cut.

Fig. 2.

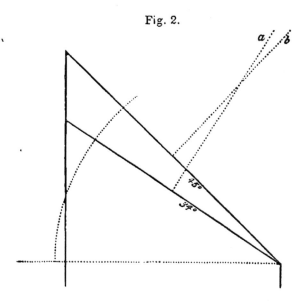

In this cut we have given the altitude of the sun at Philadelphia, a, with the roof at right angles to it, on an angle of 34°. At b, we have given the altitude of the sun at London, with its corresponding angle of elevation, 45°, and, according to the principle here laid down, both of these roofs should be equally influenced by the sun, notwithstanding the difference of his altitude at the respective places.

In a theoretical point of view these principles are correct, and are certainly preferable to the usual mode of putting on roofs without regard to anything excepting the caprice or fancy of the architect or builder. But, as general principles, we regard them as unsafe and dangerous, were they to be practically acted upon in the Southern States. Suppose, for instance, the roof should be laid at right angles to the sun in mid-summer, — as is sometimes done in England, upon this principle, — then the consequence would be that his rays would be unendurable by any species of vegetation. The mid-summer sun, even in the latitude of Baltimore, (39° 45′,) falling on a transparent surface, at right angles to the impinging rays, would scorch vegetable forms, and dry them up in a few hours.

It is, therefore, absolutely necessary that the exercise of this principle be limited to northern latitudes, where it is indispensable to economize the sun's heat, for the purpose of accelerating the maturation of fruits. It may, also, be applied in more southern latitudes, when all the warmth of the sun's rays is required early in spring; and, therefore, if the principle be applied south of the 40° of latitude, it should be taken when the sun is at its very lowest altitude, otherwise the pitch of the roof will be too flat for the months of summer.

We are decidedly of opinion — and this opinion is fully concurred in by some of the most learned and skilful gardeners in the country — that a great deal of error is committed in the pitch of hot-house roofs; probably more than four fifths of them are made too flat; their angles of elevation are much too small for the climate; and yet, notwithstanding the fierce heat of our perpendicular sun in summer, this practice is daily persisted in. One would suppose that the scorching of vine-leaves, peaches, and other plants, would convince people of the impropriety of erecting their hot-house roofs at right angles to the sun's rays in any of the summer months; and yet we know some of the finest graperies in this country on angles of about 20°. If we consider how very few of the rays are reflected by the glass, as its plane approaches a perpendicular to the sun's altitude, and how *many* are reflected as the angle of incidence is increased, we will then have some notion of the advantage of increasing the obliquity of the roof.

The annexed table will show the number of rays reflected

from various angles between the plane of the horizon, **and within** two and a half degrees of the perpendicular.

Bouguer's Table of Rays reflected from Glass.

Of 1000 incidental rays, when the angle of incidence is

87° 30', ... 584 are reflected.	60°, ... 112 are reflected
85°, ... 543 " "	50°, ... 57 " "
82° 30', ... 474 " "	40°, ... 34 " "
80°, ... 412 " "	30°, ... 27 " "
77° 30', ... 356 " "	20°, ... 25 " "
75°, ... 299 " "	10°, ... 25 " "
70°, ... 222 " "	1°, ... 25 " "
65°, ... 175 " "	

The slope of hot-house roofs, therefore, should depend on the following circumstances :

The latitude under which they are erected. — If in a southern latitude, the plane of the roof should be as oblique as possible to the sun's rays. South of 40°, the angle of incidence should not be less than 20°. It will be recollected that this angle is contained between the sun's rays and a perpendicular to the roof.

The position of the house, and the purposes for which it is intended. — Houses intended for the forcing of fruits in winter, may have their roofs made on a perpendicular to the sun's rays at that season. Conservatories attached to dwelling-houses may also have their roofs perpendicular to the rays of the winter sun, for the same purpose; but blinds should be provided for them, during the months of summer, to guard against the effects of the perpendicular rays when the sun is crossing his meridian altitude.

SECTION III.

1. *Forcing-houses, culinary houses, &c.* — Forcing-houses are erected with the intention of forming an artificial climate for the culture of tender plants and vegetables in winter and early spring. For this purpose artificial heat is employed to keep up an exciting temperature, and, therefore, it is desirable that they should be constructed in relation to this end.

Until very lately, the form in which forcing-houses were constructed was that of lean-to, or single-roofed, houses, with sheds or garden-offices on the back of them. When it is not necessary that light should be received from all sides of the house, these lean-to houses answer very well, and possess many conveniences which cannot be obtained with span-roofs. Climbing plants, such as grape-vines, trained beneath the glass, and peaches, trained in the same manner, derive a sufficiency of light from the single roof to enable them to bring their fruit to perfection; and it is very doubtful if single roofs will ever be entirely superseded for the purposes of winter forcing.

Fig. 3.

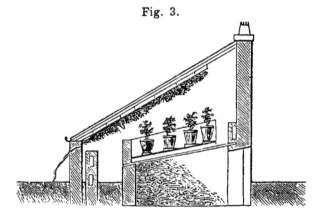

Fig. 3 is the section of a pit for winter forcing, which we consider well fitted for the several purposes to which these pits

4*

may be applied. The one here represented is what we have formerly used for the culture of grape vines, French beans, and strawberries, during winter; and where fermenting manure is to be had in abundance, it is probably the most economical house for this kind of forcing.

Fig. 4 is the plan of a forcing pit. This house is 80 feet long, in two divisions of 40 feet each. It is chiefly intended for forcing vines in pots, and is furnished with a bed, *b*, which is filled with fermenting materials for plunging the vines in, and supplying them with bottom heat. A shelf, *c*, elevated to within about 20 inches of the glass, on the back wall, and extending the whole length of the house, is intended for forcing strawberries in pots; *d* is another shelf, for the same purpose, on the front wall.

We have designed this pit with the view to procure the greatest accommodation in the given space, at the smallest expenditure for construction, keeping strictly in view the purposes for which it is intended. For winter forcing, we decidedly approve of this kind of house above all others, *i. e.*, where utility only is considered in regard to it. The cost of this house is only four hundred dollars, or eight dollars per linear foot.

A house for winter forcing should never exceed 40 feet in length, even where the operations are extensive. Thirty or 35 feet is considered, by the best gardeners, the most desirable length. If the range be a greater length, and the operations very extensive, it should be subdivided into either of the dimensions here stated, and each division heated by a separate apparatus.

There is no branch of gardening that requires a greater amount of skill, or is more calculated to display the mastership of the gardener's art, than winter forcing. It is absolute folly for any novice in gardening to attempt it. To be successful in producing the luxuries of summer, in winter or early spring, requires a great degree of skill, vigilance, constant and persevering energy. The most unwearied attention is requisite, from the day the house is started into work, until the productions are all fully matured. Scarcely a day passes but something happens, tending to thwart the object of our labors. Heat or cold, wind or

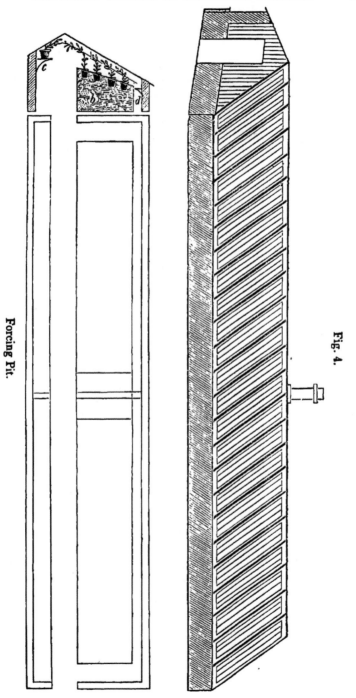

Forcing Pit.

Fig. 4.

steam, moisture and drought, mice, worms, slugs, aphides, and insects innumerable, as Cowper says, oft work dire disappointment, that admits no cure, and which no care can obviate. It is, therefore, the more requisite that the structure intended for these purposes should be the best that science and practice can adopt.

Fig. 5 is the end section of a forcing-stove, which we have seen used in various parts of this country, with considerable success. It is sunk a few feet into the ground, so that the roof reaches within about two feet of the ground level. In some places this kind of pit answers very well, as in very dry and sheltered situations. The site of such a pit must necessarily be in gravel, or sand; in wet clay, coldness and dampness would be unavoidable; and in exposed situations, it would be very unsuitable for winter forcing, unless provision were made for covering it at night.

Fig. 5.

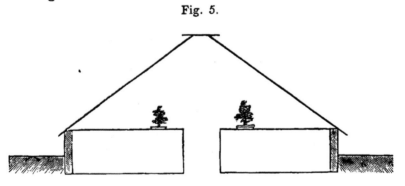

Fig. 6 shows the end section of a polyprosopic forcing-house, which, by some, is considered superior to all other forms for winter forcing. The roof presents the different faces to the sun's rays, *a, a, a,* at different periods of the year. This kind of roof may be considered as exactly equivalent to a curvilinear figure, whose curved lines shall touch all the angles of the faces, so that, were the house built in the form of a semi-ellipse, or having curved ends, the sun would be nearly perpendicular to some one of the faces every hour of the day, and every day in the year.

The rafters in this house are curved the same as in a curvi-

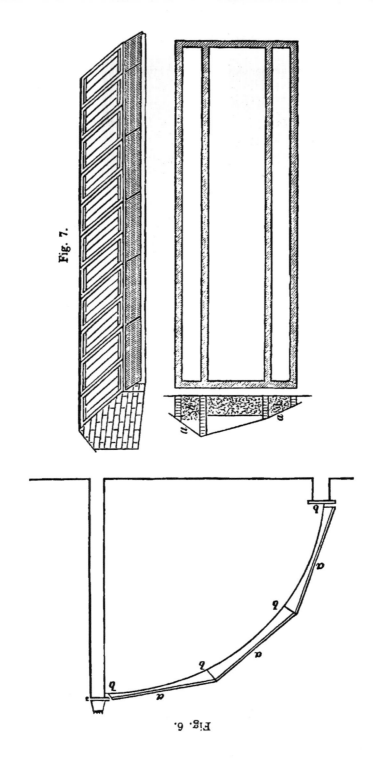

Fig. 7.

Fig. 6.

linear house, and should be made of iron, as a curvature for this purpose can be made cheaper of iron than of wood, and is tighter and more durable.   Iron beams are made to screw into the rafters, *b, b, b, b,* having a fillet in which the smaller rafters are placed, on which the sashes run.   We have seen two methods of constructing this kind of roof, — the one just described, in which the sashes are made to slide, and another, in which the sashes are made to rise on hinges, by which the house may be aired, over the whole surface of the roof, or entirely exposed, for admission of a congenial shower of rain, or for hardening the vines or peach trees, after the crop has been gathered.   The arrangement by which this is effected is exceedingly simple, not liable to get out of repair, and is applicable to all kinds of houses, whether the roof is formed of curved or straight lines.   This form of house is considered by Loudon as the *ne plus ultra* of improvement, so far as air and light are concerned.   We are of opinion, however, that these considerations alone render it less valuable in this country than it is in England, except, as we have already stated, for the purposes of winter forcing.

The Cambridge pit, Fig. 7, is admirably adapted for early forcing, where there is an abundant supply of stable manure. It is heated entirely with fermenting material, and is much used in England for the purpose of growing pine-apples, melons, cucumbers, &c.   *a, a,* are shutters, which lift entirely off, or are wrought up and down by hinges attached to the back wall of the pit.   These shutters are made to fit closely on the lining bed, *b, b,* which is kept constantly filled with the materials to supply the heat, which enters the interior of the pit through pigeon-holes in the wall.   We have kept pines during long and severe winters, in this kind of pit, keeping up a temperature of 50° to 55° in the coldest weather.   During winter the linings require to be frequently renewed, at least every week some fresh material must be added, otherwise the heat will decline below the minimum temperature ; and, as it will be some time before the new linings generate much heat, a part should only be re-newed at one time, and never both sides of the pit at once.

Saunders' forcing-pit, Fig. 8, is considered an improvement

upon the foregoing. This pit has a double roof, and is furnished with the dung-beds, *a, a,* on each side of the house. The fermenting material is supplied by means of linings along both sides of the pit, and communicates the heat to the beds through the arches in the side walls. This pit has a narrow path in the centre, which admits of the internal operations being carried on with more facility. We have only seen this pit in use by the inventor, and, so far as we know, it is quite original. Mr. Saunders informs us that it answers the purposes of early forcing better than any other construction he has tried, and works admirably, in the severest weather, without the aid of fire. We have the fact of its perfect adaptability fully verified by its productions, and are so fully satisfied with its superiority as a dung-pit, that we are about erecting one ourself. It ought to be borne in mind, regarding this pit, that unless there be abundant supplies of fermenting manure always at hand when required, it would be useless to attempt forcing with it in winter ; but this fact also applies to all forcing pits heated solely by fermenting materials.

Fig. 9 is the end section of a curvilinear-roofed cold-pit, for protecting plants not sufficiently hardy to stand the winter without protection, yet hardy enough to endure a considerable degree of cold, and even a slight frost, if kept in a dry state. Of this class we might name verbenas, roses, pansies, &c. Indeed, there are many summer flowers, used by the amateur, for the decoration of his parterre and flower-garden, which he might save, during the winter, in such a pit. The pit here given we consider the best, for any purpose to which the cold-pit can be applied. We have found them practically superior to all other pits we have yet used ; and as iron is now coming into general use, for the construction of horticultural buildings, we believe that these pits will be found, not only the most convenient, but also the cheapest that can be erected. *a*, shows the bed in which the plants are placed — we generally put in about a foot deep of tan, or saw-dust, for plunging the pots in ; — *b, b*, shows the sashes, elevated for the admission of air, supported by iron rods, *c, c*, which are made to enter a staple, by being bent, or hooked, at the end.

Fig. 9.

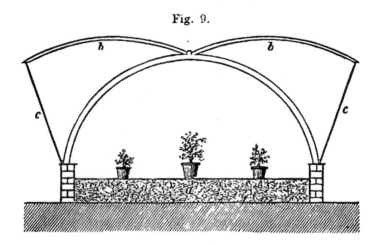

Fig. 10.

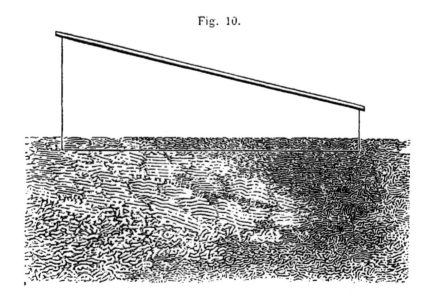

Fig. 10 is a representation of an ordinary dung-bed, with the frame set on it. The formation of dung-beds is so simple as hardly to need a single word of explanation; nevertheless, a few passing remarks may be useful to the uninitiated.

Hot-beds of fermenting materials are generally laid on the surface of the ground. Some prefer the basis of the bed to incline slightly towards the horizon; but we can see no utility whatever in this system, except the site of the bed be very wet, and then we prefer building the bed on a layer of brushwood. It is also beneficial to place a layer of brushwood every eight or ten inches deep, which lets the rank heat and steam escape more readily. The bed should have a slight inclination towards the south, when the frame is laid, though this rather tends to prevent the bed heating equally all over; and, where light is not an object, as in cutting-beds, &c., we prefer it quite level, and even inclining towards the north, the inclination of the frame turned in the same direction.

Temporary or portable frames, or cases, for covering beds, and protecting plants, are exceedingly useful about places where it is requisite to harden young plants, or protect individual specimens in the open ground.

Fig. 11 shows a portable glass frame, of a rectangular shape, and which we have often found useful for hardening young stock, in the early part of summer, which was intended for bedding out in the flower-garden. It can also be set on a dung-bed for growing early melons, cucumbers, and starting young plants into growth; for this it is admirably adapted, as the light is admissible all round.

A portable frame of this kind may be made of any size. We find, however, that about four feet wide, and six or eight feet long, is the most convenient size for practical purposes.

Fig. 12, the portable plant protector, which will be found exceedingly useful for covering individual plants, standing in the open ground. Those may be glazed with coarse glass, or covered with oil-cloth. They will be found of much utility in covering the more tender conefirs during winter, as well as during summer from the intense heat. By having the south side of the case painted with a slight coat of a lime solution, to darken

5

Fig. 11.

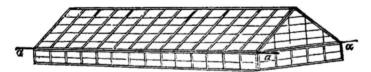

Fig. 8.

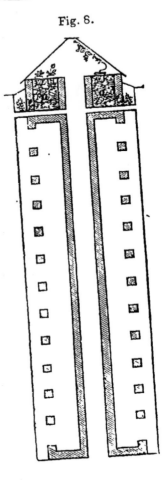

Fig. 12.

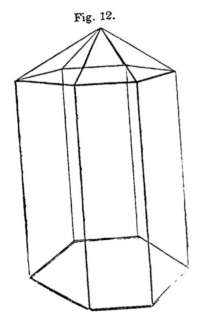

the glass and prevent the entrance of the solar rays in that direction, the plants are better able to endure the extremes of either heat or cold, than if exposed or covered with straw or mats.

In using these protectors for winter covering, it is only necessary to throw a garden mat over the case during severe frosts, removing it when the weather becomes mild, or immediately on the relaxation of the frost. There is not the slightest injury resulting from the taking off the mats, as would be the case with mat and straw coverings without the protector, as a body of air is always at rest inside, which prevents the temperature from falling so low as to cause injury to the tree.

Framing-Ground. — This term seems to have a very different meaning in American gardens from what it has in England, for we find the spot usually appropriated to the pits, frames, hot-beds, &c., located in some out-of-the-way corner, with dung, weeds, and rubbish lying about in all directions, or, perhaps, we may observe them occupying a place in one of the squares of the garden, a site equally objectionable.

Where frames and hot-beds are extensively used, they should, by all means, have a place appropriated to themselves, and sheltered, if possible, on the east, north, and west; and, as we can see no reason why this department of the garden should not be visited by the proprietor as well as any other, it should be laid out and kept in a manner to make it worthy of a visit. In fact, the frame-ground should come as naturally in the course of promenade as the larger fruit houses. Every one, indeed, may not take the same interest in this department as in others of the garden, but this can form no excuse for huddling the frames and hot-beds into some recess, out of the way, and paying no attention to order and cleanliness about them. Who, that is in the habit of frequently visiting large gardens, has not heard the gardener apologizing for the filthy condition of his frame-ground, when the curiosity or interest of the visitor led him thither? The only reason that can be given for this state of things is, that the frame-ground is seldom intended to form a prominent object in the establishment; its object being altogether for utility, it is considered by many a matter of absurdity to make it also an object of beauty.

If gardeners would consider how much gratification they sometimes lose themselves, by depriving this department of the garden of its interest by proscription, they would exert themselves more to bring it forward into its right place. If it is not a source of interest to others, it should be made so to the proprietor, for it must not be forgotten, that the pleasure and satisfaction derived even from culinary hot-beds and forcing-pits, does not wholly consist in their receiving the produce thereof, when ready for use, — for if so, recourse need only be had to the markets, — but, also, in marking the progress of their development, from the commencement to the close of their growth, in beholding fruits and vegetables flourishing in an artificial climate, and in the satisfaction of partaking of products of our own growth.

When the ground rises towards the north part of the garden, this is doubtless the most eligible site; although we are aware that some prefer placing them within an enclosure *inside* the garden, yet we think they are better placed near the northern boundary. As dung is at all times necessary, and at all times being carted to the frame yard, it is a continual nuisance having it taken over clean gravel walks. It is, above all things, desirable to have the spot approachable by carts, without in any way coming upon the gravel walks, which are appropriated only to promenade.

Fig. 13 shows the disposition of the forcing-houses, frames, etc., at a gentleman's residence in the country, which is now being executed under our direction. The ground on the north side of the garden rises somewhat abruptly from the principal range, which gives the houses a fine aspect and a dry site. Immediately behind them, and stretching along the whole length of the forcing-pits, and frame-ground, compost-ground, etc., is a belt of trees, which have been planted expressly for the purpose of sheltering the spot from the north and north-eastern winds, the same object being attained by rising ground and plantation on the west. Abundant space is left between the different erections to afford room to promenade and inspect the whole

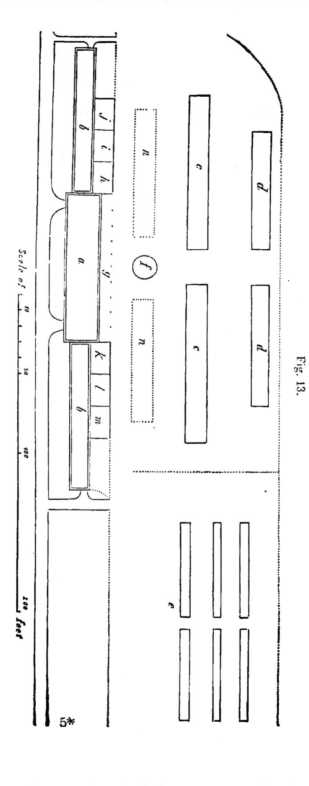

Fig. 13.

department, without being annoyed with manure under foot. Here, also, sheds and offices have been erected for the various purposes of the establishment, and arranged with a due regard to convenience and economization of labor in the operations daily going on in this department of the garden.

The position of the framing-ground should command a good supply of water; either a natural stream should be brought through it, or a plentiful supply kept in a large tank, as in the plan, Fig. 13, and kept always full for immediate use; either by means of a water-ram, or other forcing-power. Pipes should be led from this large tank or reservoir into small tanks, one of which should be in each house, to be kept at the same temperature of the atmosphere of the house in winter, for watering the plants. These tanks should receive the water from the roof, and be supplied from the reservoir, when that is exhausted.

Fig. 13 is a ground plan and arrangement of frame-ground designed by the author for a gentleman's garden.

REFERENCE TO PLAN.

a Orange house.
b b Vineries.
c c Vine-stoves for forcing in winter, **the vines being grown** in pots.
d d Culinary stoves.
e Cold frames.
f Water tank.
g Open shed for soils.
h Seed room.
i Garden office.
j Miscellaneous store room.
k Potting room.
l Store room for pots.
m Tool house.
n n Large beds, in which green-house plants are plunged in ashes during summer, being covered, during the heat of the day, with awnings fixed on **rollers,** mounted on a slight frame-work.

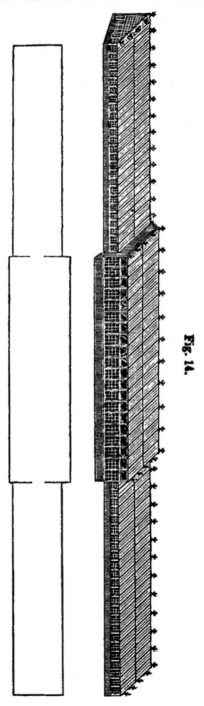

Fig. 14.

2. *Graperies, Orangeries, &c.* — These we have distinguished from forcing-houses, as not being stimulated before their natural season of growth, artificial heat being sometimes applied, however, for their protection from early frosts in spring, and for ripening the fruits or accelerating the maturation of the current year's shoots in autumn.

A greater latitude may be taken, in the construction of houses of this class, both as regards extent and ornament. Here the taste and wealth of the proprietor may be indulged to any degree. These structures may vary in length from 30 to 100 feet, or more, although we prefer them to be limited to the latter dimensions, adding others of different proportions, rather than continue the unbroken flatness of the roof beyond this extent.

Fig. 14 represents a range of houses of this class, erected by John Hopkins, Esq., in the gardens of his splendid country-seat at Clifton Park. This is one of the most extensive structures of this kind yet erected in this country. It is three hundred feet in length, by twenty-four in breadth. The structure is divided into three compartments of one hundred feet each; the centre compartment, which is larger and loftier than the others, is appropriated to the growth of orange trees planted in the ground, which, in a few years, will form a complete orchard of orange and lemon trees.

The site of these houses is one for which nature has done comparatively little, but for which art and outlay have done much, and for which the taste and munificence of the proprietor are still doing more; but, like many other structures which have come under our observation, they contain much inferior glass in the roof-sashes, which is very injurious to tender foliage. Bad glass is an abundant material in the United States, and is generally used by tradesmen, who do the work by contract, on account of its cheapness. This is a matter which demands particular attention from those erecting horticultural buildings; otherwise, they may not discover the error, until too late to prevent it.

Fig. 15 is a representation of a model house for growing grapes on the lean-to or single-roofed system; and, both in regard to its dimensions and slope of roof, is just such a struc-

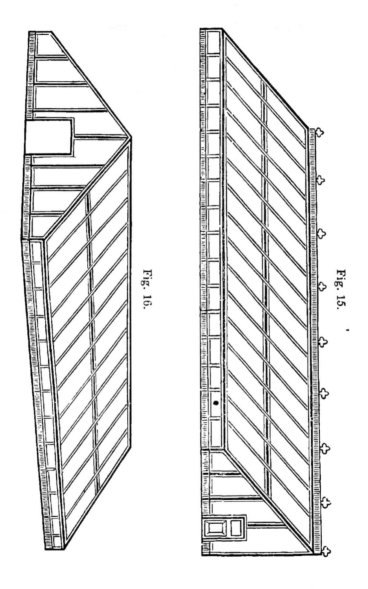

Fig. 15.

Fig. 16.

ture as we would recommend — that is, if a lean-to house was desired by the erector, or the position would not admit of any other kind. We need hardly mention that houses of this kind are suitable in many positions where curvilinear houses would be inappropriate, and where span-roofed houses would be impracticable. This house is at once cheap and substantial, in every way adapted for grape-growing, and presenting as good an appearance to the spectator as one that would cost double the sum, without any corresponding advantage.

Fig. 16 is a span-roofed house on the same scale and the same design. Of course, span-roofed houses are to be preferred, either for plant-houses or for cold vineries, to lean-to houses, although, as we have said, there are positions which render lean-to houses preferable, even as cold houses. Span-roofed houses cost somewhat more in their erection than single roofs; nevertheless, we consider it a matter of economy to erect a span-roofed house where the position is suitable, because the difference of cost is not so much as the difference of glass surface available for the growth of vines. In fact, a span-roofed house gives just two single-roofed houses of the dimensions of one of its sides. Hence, it is clear, that as many grapes can be grown in a span-roofed house, 50 feet long and 20 feet wide, as in a single-roofed house, 100 feet long and 10 feet wide, while the back wall, 100 feet in length, is saved.

From the principles we have laid down for the construction of hot-houses, in the beginning of this section, it will be apparent that double-roofed houses are in every way superior to single ones for the general purposes of horticulture, not only on account of their superior lightness, but also as regards cost of erection. And we find this fact is now becoming generally admitted, from the prevailing tendency to erect double-houses, all over the country, where the advantages of double roofs are not sacrificed to the desire of having a more imposing and extensive appearance from a single point of view.

Amongst the various forms of curvilinear houses lately brought under our notice, is that of forming the roof of the segment of a circle, which shall equal the width of the house, — a principle which we think is not generally recognized, nor do we think it

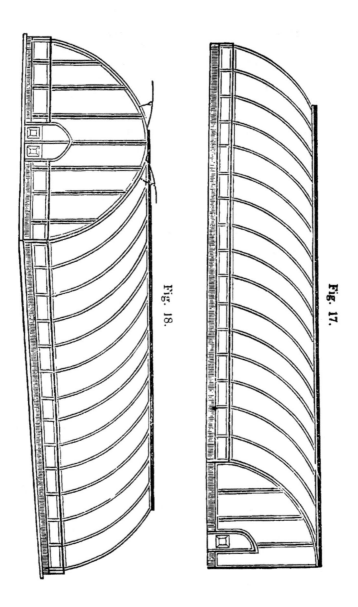

Fig. 18.

Fig. 17.

applicable except under certain circumstances. We have seen houses erected on this principle in Northern Europe, where they doubtless answer the purpose much better than houses with elliptical roofs, for the reasons already stated in regard to forcing-houses; viz., the deficiency of perpendicular light, not only in the winter and spring, but also in the early part of summer, when all the perpendicular power of the sun's rays is required for the proper maturation of the fruit. It must be evident, however, that these reasons can be of no influence on this side the Atlantic, at least, in the southern and midland states, although we know of several houses in the state of New York, built on this principle, or a very near approximation to it.

Fig. 17 is a single-roofed curvilinear house, built on the above principle, the back wall being equal to the breadth of the house. As a single-roofed house, this curve has a very good appearance, and answers admirably where perpendicular light is desirable. The only objection that can be urged against it, is the flatness in the upper portion of the roof, which gives it the same faulty character, for our hot climate, that we have urged against the flat roofs of straight-lined houses.

Fig. 18 is intended to represent a double-roofed house, on the same principle. Here the width of the house must be equal to the chord of both the sides. The parapet wall being only a continuation of the semi-circle, of course this form of house is open to the same objections as the other, (Fig. 17,) even in a greater degree, as the flat part of the roof, in this case, is precisely doubled. The perpendicularity of the rays is in some measure obstructed by a portion of the segment, at the apex of the roof, being opaque, as in the case of the house from which our sketch is taken. This plan answers the purpose very well, without depriving the house of its effect, and we think, where it is necessary, the effect might be heightened by a slight balustrade, or other ornament.

That curvilinear houses, properly constructed, are superior to those with plain roofs, can hardly be questioned on practical or scientific grounds. The construction of the monster palm house, lately erected in Kew Gardens, at London, is an evidence that this principle is recognized by the most scientific cultivators in that kingdom; and though the immense structure is avowedly for the growth of palmaceous plants, still the objections that might be urged against its modification as a palm-house might, with equal propriety, be urged against its form as a fruit house, on a smaller scale.   If there be any fault in its curvilinear construction, the fault is augmented as the dimensions of the structure are increased.

The objections that have been urged against curvilinear houses in England can have little application in this country, whatever force they might have in the cloudy climate of Northern Europe.   And we cannot help thinking that the arguments against them have, in a great degree, promoted their adoption, on account of the inconsiderate manner in which their mode of structure has been questioned.   We think it clear, that any form of curvilinear roof, from the common rectangle to the semi-ellipse, or the acuminated semi-dome, not only admits of a larger run of roof, but also a larger proportion of light, than any form of straight-lined roof that can be adopted, excepting the polyprosopic roof, which, in fact, is nothing more than an approximation to the curvilinear, or spherical roof, having the advantages of the one, without the disadvantages of the other.

Another remarkable property possessed by curvilinear roofs, and not by straight-lined ones, is their power of reflection and refraction, which, in the hot summers of our climate, is of much more importance, in a horticultural point of view, than is generally supposed.   Though the power of curved surfaces of reflecting the rays of light be similar to that of plane surfaces, yet the plane is so small on which the rays fall, that its position is changed before its concentration can cause injury to the foliage on which it falls.   As the surfaces of curvilinear roofs are, or ought to be, presented more obliquely to the sun's rays than straight-lined roofs, the amount of refraction, in very hot weather, will be greater in the former than in the latter case.   The more obliquely the ray falls on the medium of refraction, the greater the amount refracted.

The general form of curvilinear-roofed houses, in this country, is the common curvature already described, forming the segment of an ellipse, the ends being upright as in straight-lined houses.

6

For the purposes of grape-growing, we think a loss of surface is sustained by the position of the gable ends. In fact, from a series of calculations, bearing directly on this question, we have found in some houses that stand apart from other structures a loss equal to one third the extent of the roof surface. Some houses may be less, but some more, than this amount. In growing grape-vines for instance, we know that the rafters — or the sloping part of the house — is the principal area for the fruit-bearing branches of the plant. Now, supposing that your house be 50 feet in length, 15 feet wide, and as many feet high, then, by having no vines of any account growing on the ends of the house, you lose a transparent surface equal to nearly one half the extent of the whole roof. If it be asserted that the perpendicularity of the gables is necessary for the admission of horizontal light, we think this wholly unwarranted ; for experience has fully proved that horizontal light, entering by the medium of upright glass, is powerless, comparatively speaking, for assimilating the juices, either in proper quantity or quality, for the production and maturation of fine fruit. Many of the oldest and most experienced gardeners prefer hot-houses having no upright glass at all in front, placing the roof directly upon a parapet 18 or 20 inches in height.

By way of remedying the objection here pointed out, we have designed a house which combines the advantages of a curved roof with those of a plane surface, rendering the whole of the house available for the production of fruit. By this plan a greater training surface is obtained, for the same extent of glass surface, than by any other we know, or in any other structure of similar dimensions. This we consider the most perfect form of a hot-house that has yet been erected.

Fig. 19 is intended to convey a clearer notion of the kind of house we have referred to. This house is 100 feet in length, 20 feet in height at the back wall, with a perpendicular rise of five feet. The roof rises in series of successive planes, from the upright front, and presents a continuous surface for training the vines to, from one end to the other. Fig. 20 shows the ground plan of the house, which may be made of any dimensions, as easily as any of the common forms.

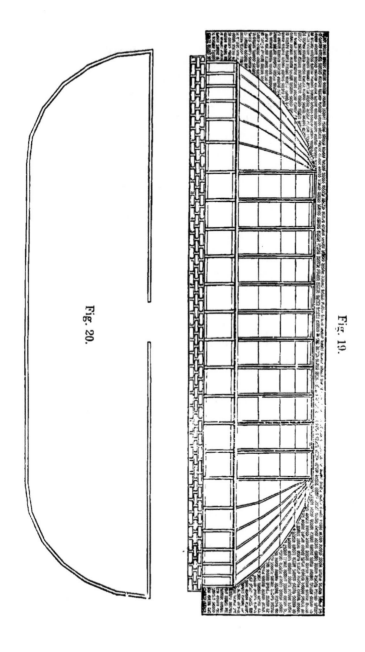

Fig. 19.

Fig. 20.

A double-roofed house can be erected on the same plan, by substituting a row of columns along the centre of the house for the support of the ridge, in place of the back wall; one of the planes being raised the necessary height at each end, for the doors, which must also be done in the single roof, (Fig. 19,) unless the door enters through the back wall, which, in some cases, may not be so convenient as having them at the ends, though, for the economizing of glass surface, we would prefer them in the back wall.

Although double-roofed houses are generally of a rectangular shape, yet they admit of every combination of form without militating against the admission of light and air. Nevertheless, that they may be perfectly adapted to the end in view, there are rules to be observed, and errors to be guarded against, which it is necessary here to point out.

If the house is above fifteen feet in width, it is necessary to have a single or double row of columns in the centre to support the ridge of the roof, but in many houses these columns are three times thicker and heavier than they ought to be, even with a due regard to strength and durability. When the columns are disproportionately heavy, the house has a dull and clumsy appearance, and the effect within is extremely bad. Indeed, columns ought to be dispensed with where they can possibly be spared, consistent with strength in the structure. We have frequently seen the internal view of double-roofed houses completely spoiled by the clumsiness of the columns supporting the roof, even when columns were altogether unnecessary. Cast-iron columns are always preferable to timber, even when the structure is made of the latter material. When the columns or rafters are bound together by braces and crossbars of slight construction, as of iron in different forms, vines and other climbing plants may be trained upon them, and be hung in festoons from column to column, or otherwise, as fancy may dictate; this gives an elegant appearance, and is always pleasing to the spectator.

Another common error in the construction of fruit-houses is, the heaviness and height of the front, something in the fashion of the heavy and dull-looking plant-houses of the last century

This results from a very general desire to give the structure a finer effect from a front view; but it must be regarded as a decided sacrifice of utility and adaptation to purpose. Making the front of graperies from eight to ten or twelve feet high, is not less objectionable than to make the roof on a level with the plane of the horizon. The sides of a hot-house should never be more than four or five feet in height. This gives the structure a more characteristic appearance, and is certainly much more fitted for the purpose in view, than upright sashes, which make the roof appear to the eye only a fraction of its real extent, whether viewed from the interior or the exterior of the structure, apart from the consideration, that the upright part of the house neither produces nor ripens the berries of grapes so well as the sloping part of the transparent surface. All structures of glass, for horticultural purposes, should have a parapet wall, from 12 to 20 inches in height, on which to rest the frame-work of the fabric; then about four feet of upright glass. This modification gives the house, whether of large or small dimensions, a neat and characteristic appearance. A span-roofed house, 24 feet wide and 16 feet high, with a five-feet front, makes a well-proportioned house, and gives about 16 feet of a run for the vines under the rafters, — the slope of the roof being upon an angle of 45°, which, as we have already said, is the best pitch for a hot-house roof for general purposes.

Until these few years, the forms of hot-houses were generally plain, flat, right-lined buildings, differing in no respect from one another than in their size and relative degrees of clumsiness. Lately, however, a great improvement has taken place in the form and construction of this class of buildings. Single-roofed houses are fast dwindling into desuetude, and right-lined houses are giving way to the more light and elegant curvilinear roofs. This is an important step in the right way; and we regard those who, laying aside their prejudices in favor of right-lined houses, adopt the curvilinear shape, as conferring a benefit on exotic horticulture as acceptable to those interested in the profession as it is creditable to themselves.

Regarding curved houses, Loudon says, — "On making a few trials, to ascertain the variety of forms which might **be** given to hot-houses by taking different segments of a sphere, I, however, soon became fully satisfied that forcing-houses, of excellent forms for almost every purpose, and of any convenient extent, might be constructed without deviating from the spherical form; and I am now perfectly confident that such houses will be erected and kept in repairs at less expense, will possess the important advantage of admitting much more light, and will be found much more durable, than such as are constructed according to the methods and forms which have hitherto been recommended."

Fig. 21 is a representation of what is called the zig-zag, or ridge-and-furrow roof, which has not, as far as we know, been very extensively adopted. There are several places in England where this method of roofing has been adopted, but principally as an experiment, or merely as the fancy of the erector. The advantage of this mode of roofing is, that the rays of the sun are presented more perpendicularly to the glass in the morning and afternoon, when they are weakest, and more obliquely to the glass at noon, when they are strongest. We doubt, however, — though the arguments we have heard urged in favor of this kind of houses be indisputable, — whether the additional expense required in their construction will be counterbalanced by the advantages gained. There is no doubt the expense of their erection militates very much against them; and, if they could be erected as cheap as plane roofs, they are decidedly superior to them for graperies, as the vine can be trained up the middle of the ridge, and, consequently, though sufficiently near the glass, the intense rays of the sun will be less injurious than under a plane roof.

The ridge-and-furrow roof may be carried out either on common plane-roofed houses, or on the curvilinear principle, though doubtless the latter is more difficult of construction, and, of course, more expensive; but we have no doubt, if the principle of constructing horticultural structures were fully understood by competent manufacturers, who had directed their attention to the details of the structures, that this, or, in fact, any other form of structure, could be made as cheap as the houses now in common use.

Fig. 21.

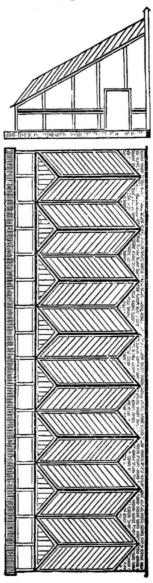

The ridge-and-furrow roof may be formed by placing the rafters as in making a common roof, say four feet apart; then placing the ridge-bars in such a manner that, contiguous to each other, they will form an angle of 45° with the furrow-bar, or rafter. Or the angle included within the ridge-bar may be formed to suit the climate of the neighborhood, — bearing in mind the principles already laid down regarding the effects of intense sunshine upon flat roofs.

The sides of the ridge may be glazed of small panes, as in common sashes, or may be made of single panes, as in the finest houses now erected; but, whichever method is adopted, the rafters should terminate in one horizontal line on the top of the parapet: this is also desirable at the back wall. Some apparent difficulty is thus occasioned in the lower part of the roof; but this difficulty is only apparent, especially if the front of the ridge be made to slope on the same angle as the side. Only the smaller and triangular pieces of glass can be used. It becomes, in fact, more economical, as the smaller pieces of glass may be all used up, which would, otherwise, be thrown away.

The ridge-and-furrow roofs are especially advantageous in countries liable to heavy falls of snow or rain, and in large houses which are parallelograms in plan. Almost any weight of snow may be carried by such roofs, especially where the furrow is small, as the pressure will then be chiefly on the bars and rafters, and not on the glass. As to hail, which is sometimes very heavy in this country, breaking the glass in flat-roofed houses, it will always meet the glass of a ridge-and-furrow house at an angle which will prevent breakage.

The advantages of these ridge-and-furrow roofs, as we have already stated, — their presenting the surface of the glass at an oblique angle to the noon-day sun, while the morning and evening sun is admitted almost perpendicular to the surface on which it falls, — ought not to be altogether overlooked in this country; and we think that a great deal might be done with houses of this kind, — probably upon an improved plan, — whereby the effect of the intense sunshine of mid-summer might be, in some measure, deprived of its meridian force upon glass-houses. Whatever may be thought of the plan here given, the principle upon which it is made is undoubtedly good ; — a principle which may easily be illustrated by placing a few common frame-sashes in the positions of the supposed ridge-and-furrow roof, placing some tender-foliaged plants beneath them, and then comparing the results, under intense sunshine, with the effects produced under a common sash, whose surface is perpendicular to the noon-day sun.

Whatever might be said in favor of cold vineries, they are, nevertheless, subject to casualties which are necessarily unavoidable. This is more especially the case in the Northern States ; and even as far south as the latitude from which we now write, (39° 45',) they are liable to the same mishaps. All houses for the production of foreign grapes should have some means or other of commanding a little artificial heat when it is found absolutely necessary. This does not amount to saying that good crops have not and may not be grown in cold-houses, without any means of raising the temperature in cold nights ; yet it cannot be denied that good crops have been sacrificed for the want of a slight fire in frosty nights. This is particularly the case in nectarine and peach houses, where we have seen the crop completely destroyed in a single night.

Experience has fully shown that the culture of exotic fruits is a precarious business, without some readily available means of averting those evils which are neither modified nor averted by any peculiar mode of construction, or any angle that can be given to the roof. This circumstance is worthy of particular attention, as many persons who design hot-houses lay particular stress on certain trifling details in the structure, which, in a practical point of view, are unworthy of the least notice.

We have lately had some conversations with men thoroughly skilled in the science, as well as the practice, of vine-growing and the details of hot-house management, and have particularly noted the diversity of opinion regarding the upright portion of the front of the house. Some are of opinion that hot-houses for the culture of fruit should have no parapet-wall, but that the sashes should rest on a water-plate level, or nearly level, with the ground, giving, as a reason, the fact that the parapet prevents the sun and light from getting to the inside border, and to the stems of vines. Now, with regard to small winter forcing-houses, this may be of some effect; but in cold summer-houses, *i. e.*, houses intended for growing peaches, grapes, etc., without fire heat, this is of no importance, as the meridian altitude of the sun during summer renders the wall rather beneficial than injurious, by shading the border during the heat of the day. Hence, it is evident that the construction of the house for grape-growing, etc., should be regulated according to the locality, as well as the period of the year at which it is required to ripen the fruit.

Many have a serious objection to upright fronts, whether of glass or other material, from the undeniable fact that fruit is seldom produced below the angle of the rafter; and if it is, it never ripens so well as that grown under the perpendicular light, nor is so well-flavored. Upright glass, however, adds so much to the appearance of this kind of building, that it can hardly be dispensed with, even at the sacrifice of a little fruit; but the latitude here allowed must be kept within certain limits, otherwise the effect produced is worse than if the house had no parapet at all.

The parapet wall of a peach-house or grapery should never be more than twenty inches or two feet high; the perpendicular sash above it, three feet more, making the upright front five feet in all. This is, we think, a proper height for structures of the kind here referred to; and this will be found to give the structure, whatever its longitudinal dimensions, better proportions, and a more handsome appearance, than if these dimensions be either diminished or increased.

In many private establishments it is much more convenient to have one, two, or more houses, than to have one single house perhaps equal to the length of the whole. We happen to know several persons who prefer erecting houses for grapes and peaches in this way; and, indeed, it has many advantages over building a large house, especially for private establishments of moderate extent, where the whole produce is consumed by the family, because one house may be advanced a month or two before the succeeding one, while the third may be protracted as late as possible, so that the fruit season will be much longer

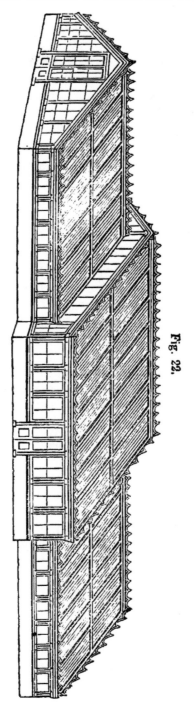

Fig. 22.

than if the structure was composed of a single house of the size of the three.

In building a range of hot-houses on these principles, say one hundred feet long, we would arrange them in the order represented in the opposite cut, Fig. 22, showing three houses united into a neat and compact range. The centre division, which is more elevated than the others, may be used as an orangery, or camellia house; or for growing figs, planting the trees in the centre bed and growing them as common dwarfs, which is the best way of growing figs, their strong and uncompliable branches being unsuited for training on the common trellises of a vinery, neither do they fruit so well as when allowed to grow like a dwarf pear-tree.

These dimensions are also advantageous on account of the trees that are to be grown in them, as different kinds of trees require different kinds of treatment, as well as different degrees of heat, air, and moisture. Each kind of tree can have the treatment which is most conducive to health and fruitfulness, without infringing on the peculiar conditions required by the others.

Where a large quantity of fruit is required, the houses for its production must, of course, be upon a larger scale. We mention this, as very absurd ideas are frequently entertained by individuals regarding the producing capacity of vines, etc., in houses, being ignorant of the quantity that healthy trees can bear without inflicting a permanent injury.

If it be desired, the centre compartment of this range may be converted into a green-house, by placing a stage along the middle of the house, and a front shelf two feet wide along the front nearly level with the building of the parapet wall, leaving a sufficient space between the shelf and the stage for a pathway. The plan of placing the green-house in the centre, between the fruit-houses, is very common. The plans of modern architects are somewhat different from those of the last century, in which we generally find the green-house a part of the culinary department, either in the middle, or in a corner of the kitchen garden. In fact, little can be said in favor of placing the green-house or plant-stove among the fruit-houses, except in small places where the limits of the ground do not admit of a select position, or where it may be desirable to place the whole of the glass structures together, either for economy, convenience, or effect.

When a grapery having some pretensions to architectural display is desired, either to correspond with buildings already on the place, or to form a connection between some portion of the mansion and another, then the structure may possess a heavier and more artistic character. This may be accomplished without in the slightest degree infringing on the principle of adaptability. For instance, there may be a recess, with the proper aspect, in some part of the mansion, which the proprietor may wish to fill up with a house productive of profit as well as pleasure; and for this purpose, he chooses a grapery, and wishes a suitable house for the purpose, without destroying the general harmony of his mansion. Or, perhaps, his premises may be very limited in extent, and he wishes a fruit-house nearly of the same order as his Tuscan or Italian villa; in which case, a house with a somewhat massive parapet and blocking-course, as in Fig. 23, would be more in unison with his taste, as well as with the rest of the premises.

This house, it will be observed, has rectangular ventilators in the front wall, which give the house a more architectural appearance; the back wall is also surmounted by an ornamental blocking-course, with ventilators for the admission of air through the back wall. [See *Ventilation.*]

We do not, by any means, justify the method of placing fruit-houses immediately contiguous to the dwelling, yet such is the taste of many. And as there is no valid reason why persons may not carry out their particular fancies with their own property, we have made the foregoing remarks for their benefit.

We do not give the above cut as a model house for an architectural vinery, — of course, its ornamental character may be increased, according to the money that is to be devoted to its erection; but with regard to the principles of its design, unless the polyprosopic roof be adopted, which is considered by some more architectural in its appearance than curvilinear roofs, when conjoined to the square forms of dwelling-houses.

7

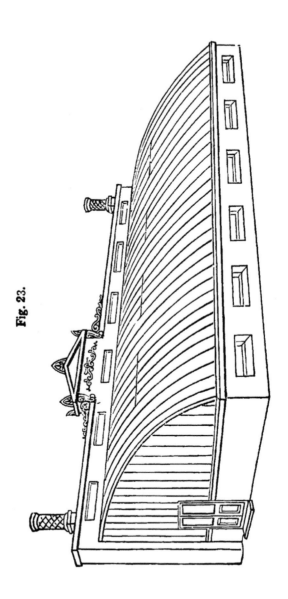

Fig. 23.

8. *Green-houses, Conservatories, &c.* — The principal distinction between a green-house and conservatory is, that in the
former, the plants are exhibited upon shelves and stages, while,
in the latter, the plants are generally planted out in a bed in the
middle of the house prepared for their reception. In many
instances, however, there is no other distinction than in the
name; as these structures are sometimes so arranged that the
middle portion is appropriated to the growth of larger plants
planted out, while the sides are surrounded with shelves for the
reception of plants in pots, as in a common green-house. And
to this arrangement there can be no special objection, especially
where the structure is of small dimensions, which admits of the
sides being shelved for plants in pots, without destroying the
character of the house, or the plants, by their distance from the
glass. We have seen a few instances, a very few, where the
two characters were amalgamated together, forming a most
interesting conjunction; but, unless the specimens exhibited be
very large and well-grown, their effect, when situated upon the
centre bed of a common-sized house, surrounded with shelves,
is meagre and defective in the last degree.

Properly speaking, a green-house is not a receptacle for large
plants, and hence it should have adequate means within it for
standing the plants within a proper distance from the glass.
This is absolutely necessary with regard to those classes of flowering plants that are fitted to adorn it, both in winter and summer. Some are of opinion that green-houses are of no further
service than merely to store away a miscellaneous assortment
of rubbish during the months of winter, for the obvious purpose
of preserving them until the next summer, that they may turn
them out under trees, or in out-of-the-way corners, to keep them
from being burnt up by the hot summer sun; and, as a matter
of course and of custom, the green-house is converted into a
lumber-room, or something else. And there it stands! what is,
or ought to be, the chief ornament of the garden, deprived of its
character, for want of taste, and divested of its interest, for lack
of skill! Visitors say, "Let us have a look at the green-
house." "No," replies the gardener, apologetically, "it's not

worth your while going in, for there is nothing there to see!
A humiliating acknowledgment, but full of truth.

It is foreign to our purpose to enter upon the present condition of green-house gardening, and the manner in which these
structures are managed by gardeners.   Our present object is to
treat of their construction, and of the means of adapting them
the most easily to the culture of flowering plants, either during
winter or summer.

It is a well known fact, that plants that are grown in what-
are called lean-to green-houses, have exactly the character of
the house in which they are grown, i. e., they are one-sided ;
nor is it possible, without a vast amount of labor and attention
on the part of the gardener, to grow them otherwise. In this
respect the cultivator does not imitate nature, but rather the
monstrosities of nature.   Trees and shrubs only grow one-sided
when their position precludes the access of light and air around
them ; but they grow naturally into a compact bush, which is
universally allowed to be the most beautiful form that plants
can assume.

Even a handful of cut flowers have their beauty, and are
generally admired, but when seen upon the living plant,
whatever shape or form the latter may possess, how much
greater their charms!   If, therefore, we add to these natural
beauties the additional charm of a positively beautiful form,
surely it will double their claim to our admiration.   And we
may here add the gratifying fact, that this claim is now generally recognized by all who can appreciate the superior beauty
of well-grown plants.

The principles upon which plant structures ought to be built,
are somewhat different from those which regulate the erection
of forcing-houses, culinary houses, &c., and as their purposes
are different, their shapes and forms are generally also different.
Plant-houses admit of a greater variety of shape and design
than any of the kinds previously mentioned, and as they are
generally erected in private grounds, for ornament and display,
they should have a more artistic character than the others.

The size of the green-house may vary according to the extent
of the collection to be cultivated, but it should always have a

.ength proportionate to its height and width. There is a great inconvenience in having the green-house very capacious, and where it is desirable to have a large collection of plants, it is best to have a conservatory for the growth of the larger specimens, or a stove for the palmaceous families of plants. We shall, however, allude to what is properly termed the greenhouse.

A first-rate green-house should be completely transparent on all sides; lean-to houses are decidedly objectionable, for the reasons already given. Houses that are only glazed in front, and have glass roofs, but otherwise opaque, are also objectionable, as plants can never be made to grow handsome. They become weakly and distorted by continually stretching towards the light, neither do they enjoy the genial rays of the morning and evening sun, and only perhaps for a few hours during midday. If such houses be large and lofty, they are still more unmanageable, as no culture can keep the plants symmetrical and of good appearance.

A green-house should stand quite detached from all other buildings, and may be of any form the fancy may dictate, or the position suggest. It may be circular, oval, hexagonal, octagonal, or a parallelogram, with circular or curved ends. The house, to be proportionate, should be about fifty feet in length by twenty in width, and fourteen feet high, above the level of its floor; if more effect be required from the external view, its parapets may be raised, to give the house a loftier appearance. The parapet should be not more than two feet high all round, the upright glass about two and a half or three feet more, including base, plate, and sash bars. The house should be surrounded by a shelf, two feet wide, level with the top of the parapet wall. This shelf is of great importance to a gardener, and is generally the best place for the finer kinds of plants; being surrounded on all sides with light, and being near the glass, they grow bushy and dwarf in habit, in which state they are most pleasing and attractive. Next to this shelf comes the pathway, three feet wide at least, (having just enough room between the roof for the tallest individual to clear the glass and rafters;) then the stage, or centre-tables, of stone or timber, and arranged

7*

according to the size of the plants to be grown.   The following
end section will illustrate what we here refer to.   It is somewhat
enlarged, for the purpose of showing the arrangements of the
interior.   The cut which follows (Fig. 25) is a perspective view
of the same house, taken at a considerable distance from it, for
the purpose of showing the effect of this plain structure in a
pleasure-ground.   If desired, it may be made to assume something of the character of a conservatory, by introducing a ground
bed in the centre, instead of the shelves or tables.   The fire-
place and heating apparatus may be placed at one end, and
under ground, so as to be out of sight, or may be formed in a
sunk shed, and blinded with shrubbery.   The flues, or pipes, for
warming the house, must be carried round, beneath the side
shelves, dipping below the level of the floor at the doors, and
returning by the opposite side of the house to the furnace.   The
cost of such a structure will very much depend upon the quality
of the workmanship, and the material used in the construction ;
but we think a very good house may be erected, according to the
foregoing plan, for about ten dollars per foot in length, or about
five hundred dollars for a house 50 feet long by 20 feet in width.

Fig. 24.

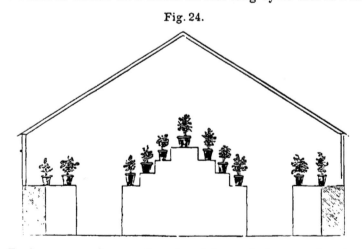

Such a green-house, though plain and inexpensive in its
character, may, nevertheless, be made to harmonize well with
flower-garden scenery, and is far superior to the clumsy, shed-
like erections frequently seen stuck into corners of buildings

and dwelling-houses, without reference to the position of the structure, or the purpose for which it was built.

Fig. 25 shows the appearance of the house, on the proportions which are given in the above plan, (Fig. 24,) which, in our opinion, admits of more room for plants than any other form that can be built at the same cost; for, although we might adopt a semi-circular form for the end toward the most prominent point of view, it must be remembered that this would add considerably to its cost. Our object here is to give the sketch of the best and cheapest kind of house that can be erected for plant-growing, and such is the one here given.

This house may be placed in any situation, as regards aspect. It may be attached at one end to any other building, without much injury to its efficiency as a plant-house; and where it is found absolutely necessary to attach green-houses to the walls of other buildings, they should, by all means, be constructed after the plan here given, or under some architectural modification of it, avoiding, if possible, that old, and now almost obsolete,

Fig. 25.

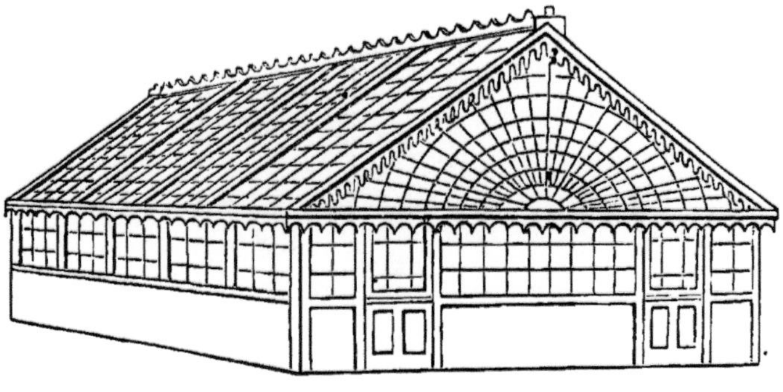

system, of laying the roof up to the wall, as in a common grapery, or of making the front of heavy pilasters and massive wood-work, like the orange-houses of the middle ages. The method of construction here described is that in which the plants enjoy the largest share of light; and this house is the easiest managed — with respect to air and heat in winter, and moisture and shade in summer — of all other methods which have come under our experience.

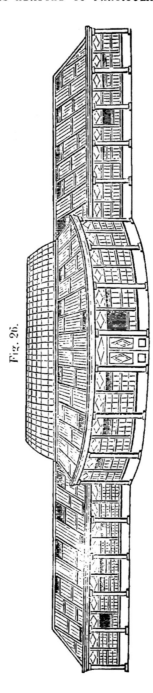

Fig. 26.

In some establishments it may be requisite to have a range of plant-houses, or one house divided into compartments, for the different kinds of plants; thus the structure may be of a highly ornamental character, as in Fig. 26, one end consisting of a common green-house, for geraniums and soft-wooded plants, and the other may be either a heathery, an orchidaceous, or an exotic stove, for promiscuous plants; the centre, being larger and more capacious than the ends, may be an orangery, or a palm-house.

This forms an elegant range of botanic hot-houses, and being of glass all round, should stand in the middle of a large pleasure-ground, or shrubbery. The smoke of the furnaces, being conducted into a subterraneous canal, is carried to a distance, and emitted by means of a shaft having the appearance of an ornamental column, as in the Botanic Gardens of Edinburgh and Kew.

By having the plant-stove in the middle of the other houses, a considerable advantage is gained by the protection afforded in winter, when the structure requires to be kept at a high temperature by artificial means; and as both of the adjoining houses will also be warmed in severe weather, the centre one, though larger, will be maintained at the required temperature with a heating apparatus no larger than the others.

From the curved disposition of the centre house, this range has a peculiarly pleasing effect, when viewed from a horizontal point of view somewhat distant. The proportions of this structure are excellent; and it would, undoubtedly, form a splendid ornament in the grounds of a gentleman's country-seat.

One of the leading errors in the erection of large plant-houses, is in the unreasonable height to which their roofs are carried, and which in the case of palm-houses may be defended as necessary; but in the case of conservatories, there is no tenable justification of such a course, except the house is intended to be the object of admiration, instead of the plants that are grown in it; and if fitness for the end in view be expressive of beauty, then, after all, these architectural temples must decidedly fail in producing that effect upon the mind, that the plain finished, but fitly and efficiently designed structure never fails to produce. But the

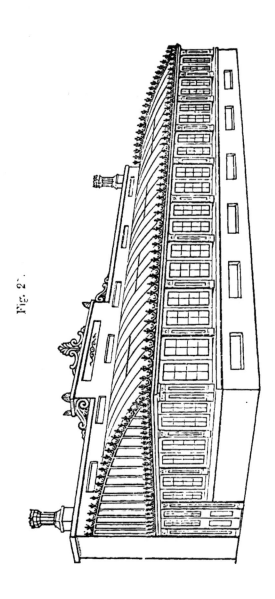

Fig. 2.

beauty, even of the plainest kind of structures, may be easily heightened and increased by an ornamental moulding of wood along the ridge of the roof, if a span, or on the end rafters and front plate, as in Figs. 26 and 27, which will deprive the house of none of its lightness, and will give it a neater and more elegant appearance.

Plants placed at a distance, either under water or under glass, are as much influenced in their development by the light as by the heat. When plants are a great distance from the roof, they are, of course, in a colder and denser medium at the surface of the soil than at the top of the house, and there cannot be a doubt that this difference in the density and temperature of the atmosphere has much to do with the struggle and effort which every plant makes to rise upward, and to elevate its assimilating organs into the warmer and most humid regions of the house. It will also be found that the difference betwixt the higher and lower strata of air in hot-houses, is more immediately the cause of plants drawing, and becoming weak, than anything that results from a feeble constitution, or from a deficiency of atmospheric air.

Notwithstanding the practical illustrations of this prevailing error in plant-houses, there seems to have been very little done to counteract this fault in lofty houses. The large conservatory in the Regent's Park, Botanic Garden, is the only structure of great size where this circumstance has had sufficient weight to induce the erectors to provide against it, in the general design and construction of the building. This admirable plant-house stands as a striking illustration of what can be done on a grand scale, without rendering fitness for the end in view subservient to architectural display, and yet, without depriving the structure of that dignity and effect which fine conservatories always convey to the cultivated mind. This conservatory, we believe, is the result of well digested practical and scientific knowledge, and we doubt if there be any other such erection in England, where the effect of this rare combination is so strikingly displayed on a scale so magnificent; and the result of this combination has indeed been clearly manifested, in the formation and subsequent management of this beautiful garden,

As the influence of the upper and lower strata of air, in large houses, will be discussed in a subsequent portion of this work, devoted to that subject, we will not enlarge further upon it at present, more than to observe, that lofty-domed, or curvilinear roofs, as that lately erected at Kew, are more difficult to manage, both in winter and summer, than low-roofed houses, whether curved or straight, and that the impossibility of rendering these houses in any way workable, has induced, in some instances, their almost entire abandonment on the part of the proprietors, owing solely to the intense heat of the superior regions of the house.

The most experienced and enlightened men have satisfied themselves, that structures in which the atmosphere has to be kept at a higher temperature than the external atmosphere, and in which plants have to be grown, should be kept at the very lowest elevation which the use and purpose will admit, so that the temperature of the air, at the level of the floor, and among the roots and lower portions of the plants, may be as little different as possible from what it is in the higher regions of the house; by regarding which, the house will be much easier kept during summer, with respect to air and moisture, and, during winter, with respect to a more equal diffusion of heat.

In the comparatively still atmosphere of a hot-house, when all is closely shut up in a cold winter's night, the difference betwixt the temperature of the atmosphere at the surface of the floor and the highest part of the roof will generally be in the ratio of one degree to every two feet of elevation; thus, in a house 20 feet high there will be a difference of 10°, and in a house 60 feet high the same rule gives a difference of no less than 30 degrees. This ratio, however, is not absolutely correct, as we have proved by experiment, in houses of various sizes, which give, under certain circumstances, a greater difference of temperature than here stated, as will be shown when we come to treat on this branch of horticultural science.*

We have already said enough on this point, here, to show the advantage of erecting low-roofed conservatories, especially when

* See Ventilation.

the object is to grow the plants in beds, or masses, irregularly placed on the level of the floor, which is decidedly an improvement upon the old method, of having a few long-legged and branchless specimens sticking their heads up to the glass, where their leaves and flowers are far above the common axis of vision, and where nothing is seen below but the monotonous bed, and the bare stems of the plants that are growing in it, compelling the gardener, at all hazard of propriety, and in violation of every principle of taste, as well as of his own judgment, to stick in the commonest plants, whatever they are, among the bare stems of the others, to fill up the unsightly blanks and vacancies thus occasioned in the beds.

While on this subject, we will just briefly remark, that nothing has so much tended to improve the culture of the trees and shrubs, generally grown in houses of glass, as the improvement that has taken place in the mode of construction. All practical men are agreed on the point, that, to grow plants well, the house must be low in the roof, and light as well as air must be admitted freely to every part of the plant, from the ground to the glass. They must also be situated in such a way, regarding their lower parts, that the light may not be obstructed, for however powerful, and perhaps sometimes injurious, the fierce rays of the mid-day sun may be in mid-summer, yet its permanent obstruction is far more so. It is easier to obviate scorching in the one case, than etiolation in the other.

SECTION IV.

1. ARRANGEMENTS for the interior of forcing-houses, culinary-houses, &c., are generally very much alike, consisting chiefly of trellises of wood, or of wire, to which the trees are trained. The other portions of interior detail are common to horticultural structures of every description, and will be subsequently described in their respective places.

"Half the advantages," says Loudon, (Ency. of Gard.,) "of culture, in forcing-houses, would be lost without the use of trellises. On these the branches are readily spread out to the sun, of whose influence every branch, and every twig, and every leaf, partake alike; whereas, were they left to grow as standards, unless the house were glass on all sides, only the extremities of the shoots would enjoy sufficient light. The advantages, in respect of air, water, pruning, and other parts of culture, are equally in favor of trellises, independently, altogether, of the influence which proper training has upon fruit-trees, as the vine, the peach, apricot, &c., to produce fruitfulness."

Notwithstanding the obvious utility of trellises in culinary houses, the use of them is frequently carried to a most unprofitable and injurious extent, when the whole interior of the house is filled with foliage from the glass to the floor. Here, work is entailed upon the gardener to no purpose; and though good crops may be borne on the trees that are trained upon the trellises crossing the house, or on the back wall, the fruit is utterly worthless.

The trellis, situated on the back wall, was formerly considered the principal part of the house, for producing a crop; but this is only the case in small, narrow houses, and where no trees are trained upon the rafters, or under the glass. Experience has proved that, where the whole surface of the glass is covered with foliage, there is very little gained by training either peaches or vines on the back wall.

The principal use to which back-wall trellises may be profitably turned, is for the cultivation of figs, which are found to do much better than peaches under the shade of others.

The trellis, whatever its form, should be as near to the glass as possible, and placed so as to command the full influence of the light entering the house. When the vines are trained upon the single rafter trellis, Fig. 28, A, leaving the middle of the lights open, for the free admission of light to plants beneath, then the curvilinear trellis may be introduced into the centre of the house, as represented at *a*, Fig. 29, from which good peaches and nectarines may be obtained, providing the sashes be kept open in the middle, as already stated, for the admission of the unobstructed light.

Fig. 28.

A.   B.   C.

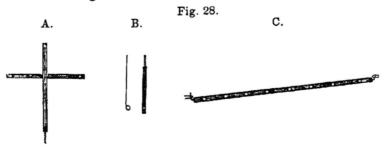

The most common method of fixing the roof trellis is by studs, Fig. 28, B, screwed into the rafter, about eight inches distant. Each stud is provided with an eye, or hole, at the extremity, through which the wire is passed, and tightened at both ends by screws and nuts. The studs should not be less than twelve inches in length, so as to afford room for the foliage to expand itself fully, without coming in contact with the glass, which, when moistened with the condensed vapor, is apt to scald the leaves that happen to be touching it. The wires forming the trellis are stretched horizontally from both ends of the roof, at about nine inches distant.

Instead of studs screwed into the rafter, the horizontal wires may be fixed, and kept in their places, by rods of iron, having holes for the wires passing through, at regular distances. These

rods are attached by a loop and staple to the front wall at the
lower end, and to the back wall at the upper. This method is
preferable to having the studs screwed into the rafter, as they
can be easily removed, or the whole tegument of trellis may,
if desired, be taken down and put up again without much
trouble. This is of great importance on occasions of cleaning
and painting the sashes, etc. Fig. 28, C, shows the perforated rod
which is here referred to, the looped end being fixed in common
staples.

When provision is made for a middle trellis, this should
always have a curvilinear shape, as in *a*, Fig. 29. This form

Fig. 29.

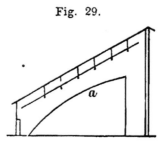

affords not only the largest training surface, but presents a
larger surface to the light, than any other form that can be
adopted, and, what is of more importance in regard to small
houses, it occupies less room in proportion to its training surface
than any other trellis with which we are acquainted.

Cross-trellises, or horizontal upright trellises in the middle of
the house, not only destroy the effect within, but are worse than
useless. Where the house is of sufficient size to admit of a
middle trellis, and a sufficiency of roof-surface to afford the centre of the sashes to be kept clear of foliage, we should prefer
having a sloping trellis on the back wall, and the centre bed
occupied with dwarf standards, planted either in a straight or
zig-zag line along the border, which, under good management,
will be as fruitful as if trained on a trellis, while their appearance would be pleasing and handsome. Fig. 29 will convey a
better idea of our method than by description. Fig. 30 shows
the same system carried out in a double-roofed house.

Trellises are now made generally of wire, as being cheaper

Fig. 30.

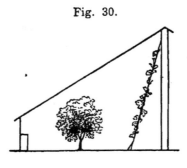

Fig. 31.

and lighter than wood. Wire is in every way fitter for tne purpose than wood, especially for roof trellising. The distance at which the wires should be placed apart depends upon the kind of trees to be trained to them. For grapes, the distance should be 12 or 14 inches; and for peaches, nectarines, and small-wooded trees, not more than 8 inches. The distance of the wires of the roof trellis from the glass should not be less than one foot for grapes, and for peaches and other similar trees not less than ten inches. In properly constructed houses, there should always be a lower trellis, with the wires placed at double the distance of the others, for training the summer shoots to, to prevent the crowding of the vine branches when the trees are full of fruit, in order that there may not be a confusion of fruit and foliage. Vines, or, indeed, any other kind of fruit trees, should never be nailed to the wood of the house; but, in all cases, trained at some distance from it, however little room there may be for that purpose.

2. The interior of the green-house is generally provided with a stage in the centre, and shelves round the sides on which the

8*

plants are arranged; and this is the principal object which demands our attention. In single-roofed houses, the stage gen erally rises towards the back wall; but in span-roofed houses, which are surrounded by a path, the stage or platform rises from both sides, and meets in the middle of the house.

It is a principle with some people to place the stage on the same angle as the roof, *i. e.*, each shelf rising at an equal distance from the plane of the rafters. This, however, is a bad rule, and, in cases where the roof is very steep, will make a wretched receptacle for green-house plants. No general rules can be laid down for the erection of the stage, as this will very much depend upon the form and size of the house. We might add, however, that the angle of the stage ought never to exceed the angle of the roof, but, if practicable, should be rather flatter than otherwise, to admit of larger plants being placed on the upper shelves, which serve to give the house a larger and more effective appearance from the inside view.

Green-houses intended for the growth of a promiscuous collection of plants, some of which may reach a considerable height, should have but few shelves on the platform, say three or four rises are quite sufficient, leaving the upper shelves, at least, twice the width of the others. This applies, also, to single-roofed houses. Many commit an error in making their stages not only too steep, but the shelves too narrow and too high, individually. The shelves of a green-house for displaying plants ought not to be less than one foot in width, this width increasing towards the top shelf, and not more than eight or nine inches in height from each other.

Houses appropriated to the growth of small plants, as nursery-men's stock-houses, propagating, etc., may be staged much closer than this. These remarks chiefly apply to the greenhouses of private individuals, and houses for the exhibition and arrangement of a general collection of plants.

3. Conservatories, orangeries, and houses for the growth of the palm family, have pits, or more properly beds, in which plants are planted out. These beds are sometimes level with the floor, and sometimes raised above it, being enclosed by a

curb. The principles of culture in these houses being somewhat different from the common green-house, it is necessary
that they be arranged to suit the plants grown in them.

The general form of conservatory beds is exactly that of the
structure. If the house be a parallelogram, the bed has the
same form, sometimes divided in the middle by a path, and
sometimes surrounded by a path on both sides. These structures,
when properly built and managed, are undoubtedly the means
of conferring on lovers of gardening and flowers, enjoyment of
the highest and purest character. When a fine conservatory of
this kind is attached to the mansion house, or connected with it
by a glazed arcade, it forms one of the most delightful prome-
nades in winter that wealth and taste can command.

There is undoubtedly much yet to be done in the way of
improving the interior of ornamental conservatories, not only as
regards their adaptability to plant culture, but also their general
effect. We seldom see anything else than the same flat, formal
bed or border, which is either rectangular, round, or square,
according as the form of the building may determine by its
walls. Even the refinement or elegancies of construction of
architecture fail to invest such buildings with any character
of distinctness or novelty, owing to the sameness or monotony
which forms the basis of the design. As far as relates to the
exterior, a considerable improvement is taking place from the
use of curvilinear roofs, and lighter and more elegant workmanship, and also resulting from the adoption of double-roofed
houses, instead of the dark, dull, narrow, clumsy shed-like erections which formerly used to be erected, and the various forms
of elevation, which are now so generally arranged as to produce
a very pleasing and picturesque effect.

A recent and very general improvement in the construction
of green-houses, consists in making the stages and shelves of
slate, or thin plates of stone; this practice is now common about
London. These slates are frequently grooved or hollowed out
so that the water is retained under the pots, and thus dripping
is prevented, and evaporation is provided for in dry weather.
This may be considered as a real improvement, which is proved
by the readiness with which this practice was adopted by prac-

tical gardeners and nurserymen, and, from the cool nature of that material, deserves to be more extensively followed in ornamental green-houses in this country.

The irregular method of laying out the interior of conservatories, which promises to subvert the formal and monotonous arrangements of the old school, is one of the greatest steps towards a higher and more natural taste of artificial gardening than any other that has taken place in this department of the art for the last fifty years, inasmuch as it can be carried out with equal advantage on a large, as well as on a small, scale; and where this method is applied to a large structure, *i. e.*, a structure covering a large area of ground, it necessarily leads to the adoption of interior arrangements, as far surpassing the old method in beauty and effect as it does in respect to economy, convenience, and comfort.

When we visit a conservatory lately erected, and see it to be a perfect fac simile of others that had been erected a century before, there is positively nothing to strike us with admiration, except, perhaps, the character of its architecture. When we see, in the costly erection before us, the exact image of conservatories everywhere else, the object loses one half of the charms of novelty and interest. It is, in fact, in the endless variety and intrinsic beauty of which they easily admit, that their chief fascination rests. This is the case with all other objects of art, with private mansions, for instance. How monotonous and tiresome would a country or suburb be, were every mansion and dwelling an exact copy of the other! And why should it be so with erections for the growth of plants? Why should these, which are, to a certain extent, invested with the charm of rarity, be deprived of the charm of variety? Why should there not be groves, and lakes, and irregular flower beds, and rocks, and aquariums, and caverns, and jets, and waterfalls within as well as without? In the former case, their beauties would be available, either for recreation, admiration, or study, at all seasons; in the latter, the fickleness and vicissitudes of our climate frequently prevent the enjoyment of either.

The finest illustration of this system with which we are acquainted, is in the beautiful conservatory of the Ro

Botanic Society's Garden, in the Regent's Park, by Mr. Marnock, and which is, perhaps, one of the best adapted structures for the growth of plants in England, and is decidedly superior to the many monster plant-houses lately erected in that country. We have compared this structure with the large houses at Chatsworth, Kew, Sion House, and other places, and, whether in respect to convenience and comfort, general appearance or adaptability, we consider it in every way preferable to any other structure of the kind we have seen. This splendid winter-garden — for its great size justly entitles it to this name — contains collections of different degrees of hardiness, and embraces climates suitable to each. Its walks are gravelled, like a flower-garden, winding through amongst the various groups of plants; sometimes overhung with the pendulous branches of flowering plants of great size and beauty, and sometimes winding beneath arches and arbors of climbers in wild profusion. Here you climb over rocks, covered with characteristic plants, and there you descend into the humid recesses of orchids and aquatics. This house has not the domed and lofty character of some other structures of the kind, which is at once a prominent feature and a prominent fault in their construction; it consists of several spans, supported on light iron columns, the centre one being somewhat higher than the others; and, though having little pretensions to what is generally called architectural display, yet its commanding position and its magnitude strike the observer with a feeling of admiration, which is only surpassed by its internal arrangements.

The general system of building conservatories in a recess of the mansion is entirely subversive of this method of internal arrangement, because of their total inadaptability for this purpose. It must not be supposed, however, that there is any absolute reason for detaching the conservatory from the mansion, if it be otherwise desired; but it ought to be there as a positive part of the building, not a tributary attachment to fill up a corner. That these kinds of structures for plants are being rapidly improved, is evident, and this, indeed, must be the case, since the improvement here spoken of springs from necessity. The attachment of a green-house to a mansion appears to us in a' questionable taste, as placing the conservatory in the middle of the kitchen garden, or in the orchard; and if any kind of horticultural structure is to be attached to the mansion, it ought by all means, to be a conservatory.

As an illustration that conservatories may form prominent portions of a mansion, or even a whole wing of it, without destroying its architectural character, we might point to a design, in the December number of the "Horticulturist" for 1849, by A. J. Downing, Esq., of Newburgh, which is introduced to show how a simple structure of this kind ought to be treated so as to give the whole an architectural and harmonious character, and showing, also, how this may be accomplished without rendering the conservatory opaque on either side, except the one end by which it is attached to the house,—a circumstance which will be indispensable in conservatories attached to houses, unless they be joined by means of a veranda, which gives them somewhat of an isolated character. This house which we have referred to is the kind of conservatory which we like, being satisfied, from experience, that, unless they be constructed somewhat after this method, they can never give the proprietors that satisfaction which they have a right to expect; and we trust Mr. Downing will go on with creations of this kind, till these transparent conservatories become more general than they are at present.

Although it is not necessary, on account of perfect adaptability, to place conservatories apart from dwelling-houses, yet we generally find that structures, standing detached from the mansion, are better suited for the growth of plants : first, because there is less temptation to introduce massive workmanship, on purpose to harmonize with the house; and, secondly, there is, in most instances, more facility of making the house to satisfy the requirements of vegetation, and, consequently, less likelihood of departing from the principles of erection which science and practice have determined as essential to the successful cultivation of plants.

In many instances, it is absolutely impossible to comply with these principles, whatever interior arrangements may be adopted. Where the conservatory is a mere lean-to, stuck-in attachment, compliance with the principle of plant-culture, or with the methoa of interior arrangement which we have here recom- mended, is equally impossible. In the latter case, the greater portion of the plant-house must necessarily be formed by the walls of the building, and the shadow of its elevated parts will be thrown upon the plant-house for at least one half the day. This is nearly as injurious as if the portions thus shaded were opaque. The only way of obviating the evils consequent upon its position, is to give every possible inch of light to the one, to enable it to counterbalance the shade which it must bear from the other.

When plants are planted in beds in the conservatory, they require to be large specimens, otherwise they have a meagre appearance, and must be a great distance from the roof, and this is one of the greatest difficulties the gardener has to contend with. It must be borne in mind that fine specimens do not consist in plants that reach from the bed to the glass, with naked stems, and only a few branches at the top, which is invariably the result of lofty roofs and dark walls.

We have already shown, in the preceding section, the conse- quence of high-roofed houses, and the difficulty of managing them in a manner fitted for the successful cultivation of plants; and if high-domed or right-lined roofs be improper in houses where the plants are elevated on shelves and stages, they are much more so where the plants are set in the beds without pots, as the distance from the light renders it impossible for them to grow bushy and branching below. These, when included within the common-place curb of a square, or a parallelogram, or an oval, or circle, which are little better, (except when sparingly introduced, and only where they are described by the natural curves of the contiguous figures,) invariably produce an effect so common-place and uninteresting, as to fail in exciting the faintest emotions of pleasure, or novelty, or interest, in one out of a hundred individuals of taste and judgment.

REFERENCE TO FIG. 32.

A, A, A, A, A, A, Beds in which the plants are set out and arranged according to their methods of growth, habits, height, &c.

B, Water Tank, with jet in the centre. This tank is surrounded by rock-work and characteristic plants.

C, C, Seats on each side of the jet, commanding, also, views of the surrounding grounds.

D, D, D, D, Conduit for the hot-water pipes, for warming the structure. This open conduit passes along the wall the whole length and breadth of the house, and is covered with grating, which serves as a path for watering, and conducting the necessary operations connected with the culture of the plants.

E, E, E, an open Balcony, passing all round the house, and surrounded by a balustrade. This balcony forms a continuation of the porch on the one side, and runs out upon the ground-level on the other. From this balcony are seen the garden, the lakes, the hot-house, and the ornamental grounds. The chief purpose of this balcony, however, is to maintain the ground-level of the floor, and to make the conservatory in harmony with the mansion, without destroying its adaptability as a first-rate plant-house, of that class intended for growing large specimens, planted out in the ground.

F, Steps, leading from the balcony into the pleasure-grounds.

G, Door opening from the drawing-room.

H, Rock-work for alpine plants, surrounding the aquarium and jet.

For end view of this house, see Frontispiece.

Fig. 32.

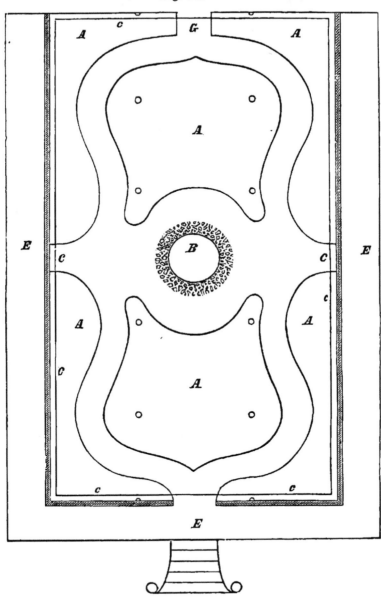

Conservatories are, probably, the most important structures used in ornamental gardening; and, as we have already said in regard to other kinds of horticultural buildings, we say, also, of them, that no degree of gardening ability, and practical attention on the part of the gardener, will compensate for the want of light and air; and, where the arrangements for the working of the house, in regard to air, heat, &c., are imperfect, the risk is great, and it is painful for a skilful and zealous gardener to contemplate the consequences which he may be unable to prevent. One single night may destroy the labors of years past, and forbid hope for years to come; and, after all, the blame may be laid where it is least merited, and censure withheld from the party who most deserved it.

In all buildings, and especially conservatories, the most complete and elegant design, when badly executed, is disagreeable to the view, defective in the object of its erection, and ruinous to the proprietor, because it is incapable of giving that satisfaction and pleasure which he was entitled to expect from his outlay.

Fig. 32 is the ground plan of a conservatory, which we have designed for erection at a gentleman's country-seat. It is intended to form a prominent wing of the mansion. The structure is entered at one end by a door, leading from the principal apartments of the house. The conservatory is traversed by curved walks, laid with marble, and bordered by a curb, on each side, of the same material. In the centre is a basin of water, with a jet playing over a rockery, as seen in the cut, Fig. 32. In this design we have endeavored to combine perfect adaptability, with beauty in the structure, and harmony in the whole.

This method of laying out the interior of a conservatory admits of the most perfect arrangement in the planting of the beds and compartments, intended for the exotic trees and shrubs, with which the structure is to be filled. The walks wind through, among the plants, as in a common shrubbery, or flower-garden; and, when the compartments are tastefully arranged, and the whole kept in healthiness and luxuriance, with climbing plants hanging in festoons from the rafters and other supporters of the roof, it forms decidedly the most delightful and satisfactory kind of horticultural structure that can be erected for comfort, convenience, and enjoyment.

We do not think that any definite rule can be laid down for the laying out of the area of a conservatory, as the formation of the beds and walks may be dictated by the taste of the proprietor, or those in whom he confides the management of the work. Almost any curve may be adopted in the walks, without destroying the effect of the interior view. What we condemn is the monotonous straight lines by which the area is generally laid out. It must be observed, however, that this method is entirely inapplicable, unless the house be glazed on at least three sides, and the roof so constructed as to admit the greatest possible quantity of light in proportion to the extent of the area enclosed. The roof should, also, be as low as is consistent with exterior effect, and the admission of plants of good size; for, as we have already observed, one of the prevailing errors in the construction of conservatories adjoining mansions consists in their being made too lofty and too opaque. They are designed generally to suit the place of the building, without regard to the effect of the conservatory itself, as a structure, or as a plant-house.

There are many other advantages, resulting from houses of this description, which, in a practical point of view, are deserving of consideration. Not the least of these is the facility with which plants can be arranged to produce the best possible effect. Plants are much easier arranged within curved lines, than in squares or parallelograms; and the curvatures of the beds are always more spirited and pleasing than continuous straight lines, whatever the house may be filled with, or however badly the plants may be disposed.

We have only room to notice one feature more in the construction of this conservatory, viz., the form of the roof. We have chosen the spans of different sizes, in preference to one single span, as much for adaptability as to harmonize with the architecture of the mansion. This system tends to prevent the accumulation of warm air at the top of the house, and hence the heat is distributed more equally among the plants. For the same reason, ventilators are provided at the top of each span, so that the external air admitted, as well as the artificial heat rising upwards, will be more equally distributed over the house.*

* For further notice of this, see *Ventilation.*

An end view of this structure is shown in the frontispiece. As
the ground, in this case, descends gradually from the base of the
mansion, a considerable depth of parapet wall is necessary to
bring the floor of the conservatory to the desired level, and the
requisite distance from the roof. Curved roofs can only be
adopted where the building admits them without jarring dis-
cordantly with the general architecture, and, in some instances,
straight-lined roofs will be preferable; but in all cases where
curvilinear roofs can be made to harmonize with the building,
they are decidedly to be preferred, on account of the superior
beauty of curved lines viewed in contrast with the surrounding
scenery, and also on account of the superior beauty of the structure from within, in harmony with curved figures of the walks
and borders of a house, such as that we have here described.

SECTION V.

1. *Workmanship.* — However excellent and adaptable may be the design of a horticultural erection, if the work be badly executed the structure will generally be defective in the working, and the trouble of management will be greatly increased. Bad foundations, bad roofs, bad-fitting sashes, rendering them difficult to open and shut, bad glazing, and bad workmanship of every description, are too common to exist without being a very perceptible evil, and one that is much complained of by practical gardeners, upon whom the consequences of this method of construction generally fall. In all regular work, coming under the province of the architect or engineer, there is generally particular attention directed to the facility of working, and ingenuity is exerted to its utmost limits to perfect and simplify those facilities, however temporarily the structure or work may be constructed. But horticultural buildings, relatively to civil architecture, appear to be an anomalous class of structures, not coming strictly within the province of the architect, — except in so far as they may be related to the house in an architectural point of view, — and hence they are more the subject of chance or caprice in design, and of local convenience in execution, than any other department of rural architecture. The subject of horticultural architecture has not been deemed of sufficient importance to induce civil architects to make themselves acquainted with the principles on which plant-houses should be constructed, or to consider the nature of workmanship in relation to its work; and, consequently, the construction of horticultural buildings is either left wholly to gardeners, who understand little of the science of architecture, or wholly to architects, who understand as little of the science of horticulture. The consequence, in either case, is generally incongruity in appearance;

9*

want of success in the useful results, and want of permanency in the structure itself. In every country, no doubt, such cases are numerous, but here, they are more numerous probably than in any other, arising, no doubt, from that want of attention to the details of horticultural architecture, and to the still undeveloped principles of science, upon which it is based.

The temporary and inferior character of the workmanship generally bestowed on horticultural erections is a source of great loss to those erecting such buildings, and demands the serious attention of all who contemplate the construction of them. The remarks, which have been applied by a popular writer on farming in regard to farm-buildings, are still more applicable to buildings for the purposes of horticulture.* Buildings, manifestly intended to be permanent, are put up to stand for a year or two, when it becomes absolutely necessary to their continuation, to spend a sum upon them equal to one third the cost of their original erection, which acts as a drawback upon the progress of horticulture in this country, as many suppose that this early additional expenditure is merely the consequence which the common tear and wear of time entails upon all such structures; and hence they are considered too expensive to keep in order, even though willing to go to the cost of original construction. Now experience has taught us that structures, substantially constructed at the first, and of good materials, will stand for at least twenty years without any additional outlay, save a few coats of paint during that period, which increases their durability, the oftener it is applied.

We have been induced to dwell longer on the subject of workmanship, from the numerous examples which have come under our own observation, and from the trouble and annoyance to which we are almost daily subjected on this account. In small erections, the inconveniences arising from bad workman-

---

* Few things serve better to distinguish the habits, and even the characters, of the progeny from the parent stock, — the Americans from their English ancestors, — than the more perfect and durable character of all *their* mechanical works, machinery, and buildings. There, things are made to endure ; here, they are made to answer the purposes of the day. — [*Ed. Farmer's Library.*]

ship may be little experienced; but where the structures are large and extensive, the results become of the deepest importance, in an economical point of view.

It is not easy to point out a course wherein these difficulties may be avoided, or to discover, at all times, to whom blame is attributable. Tradesmen, who take the work by contract, probably endeavor to do the best they can with the job they have taken in hand, and it is generally their policy to get over it as easily and as quickly as possible. Gardeners who may have the superintendence of the work, probably do the best they can, but from their wanting the necessary knowledge of the details of construction, are unable to exercise that surveillance which is necessary to the proper execution of the work.

2. *Materials of the Frame of the Building, &c.* — The most suitable material for the frames of horticultural buildings has lately been made the subject of considerable discussion and experiment, which has not been without its use in the elucidation of facts hitherto unknown, or, at least, unnoticed in general practice. The case of wood versus iron has been investigated on various grounds, by practical and scientific men, without, however, coming to a unanimous decision on the superiority of either. In this matter, as in some others like itself, some have adopted extreme views of the various merits and defects of the different materials, and have come to their conclusions by reference to some single or specific property. These views and conclusions, however, have been of considerable utility in bringing the subject before the bar of unbiased inquiry, which, if it has not already done so, is likely to result in the adoption of modified views, and the recognition of specific principles, that, when fully considered and duly weighed against each other, will ultimately lead to a more definite result.

The use of iron in the construction of hot-houses, like every other really valuable improvement, has met with much opposition from the still slumbering spirit of prejudice, which is generally slow to believe in the superiority of anything different from that with which it has been long acquainted, even when this superiority cannot, on reasonable grounds, be denied. This spirit, however, which has long held undisputed sovereignty over the minds of gardeners, is fast giving way before the sweeping current of mechanical inventions; and when science comes to the aid of mechanism in the building of hot-houses, as in the erection of factories, steam-engines, and other works of art, then the flimsy barriers reared by prejudice will be swept away, and I think I may fearlessly assert that, in regard to the opposition that has been given to the erection of iron hot-houses, this has nearly taken place.

Gardeners, from the early ages of Abercrombie and Nicol, have been prejudiced against metallic hot-houses, and, to our knowledge, this prejudice is still entertained by some whose learning and intelligence would encourage us to look for more accurate judgment.

The objections which have been raised against metallic houses for horticultural purposes, are chiefly the following: —

Contraction and expansion, oxydation, abduction of heat, attraction of electricity, and original cost.

In regard to the first, and principal cause of opposition, viz., its susceptibility to the influences of heat and cold, a fact which cannot be denied, yet it is proved by experience that if a house be properly constructed of good material, this susceptibility is of no practical importance. In very small houses the inconvenience occasioned by sudden fluctuations of temperature may be more sensibly felt, although, in the management of small iron vineries, in England, we have never seen the slightest inconvenience result from external changes; indeed, all our experience in the management of hot-houses goes to prove the superiority of iron over wood, for every purpose to which timber is generally applied. It has been stated that metallic roofs are more liable to break the glass than wood; practice has also proved that this statement is without foundation, and if it has ever taken place, can only be in copper or compound metallic roofs. Cast-iron or solid wrought-iron bars have never been known to cause breakage of glass, or displacement of joints, and some have asserted that the breakage of glass is even more during sudden changes, by wood than by iron roofs.

The expansibility of copper being greater than that of iron, in the proportion of 95 to 60, therefore copper is above one third more likely to break glass than iron. But when it is considered that a rod of copper expands only $\frac{1}{100000}$ part of its length with every degree of heat, and that iron only expands $\frac{1}{166666}$ part, the practical effects of even the hottest portion of our climate on these metals can never amount to a sum equal to the expansion required for the breakage of glass.

The second objection which we have mentioned is also undeniable. All metals are liable to rust; but painting easily rids us of this objection, at least it will so far prevent it as to form hardly any objection.

The power of metals to conduct heat is an objection which, like the others, cannot be denied, but may be partially obviated. The abduction of heat, like the expansibility of metallic roofs, is very little felt in using them; the smaller the bars, the less their power of conduction. The paint, also, and the putty used to retain the glass, obviate this objection. Heat may be supplied by art, but light, the grand advantage gained by metallic bars, cannot, by any human means, be supplied but by transparency of roof.

The objection raised on the ground of attraction of electricity, is easily answered. If metallic hot-houses and conservatories attract electricity, they also conduct it to the ground, so that it can do them no harm. What is corroborative of this position is the fact, that no instance has come under our knowledge of iron hot-houses having been injured by the electric fluid.

The objection regarding the expense of iron hot-houses, has been sufficiently refuted in England, and we have observed, with pleasure, a refutation of the same objection, by an enterprising gentleman of Cincinnati, who has lately erected an iron-roofed vinery. Mr. Resorr has given a cut, and description of this house, in the "Horticulturist" for Sept. 1849, p. 117. This is the only substantial account we have seen of the comparative cost of iron and wood roofs. This gentleman, who is in the foundery business, has every opportunity of knowing the accurate cost of such a house, and plainly states, "that those wishing to build a good, substantial house, can do it, and make the roof of iron, as cheaply as of wood, the other parts costing the same." From inquiries and calculations which we have made, we have come to the same conclusion, although, from a want of the requisite knowledge, and from the expense of having patterns made for the castings, it may, in some localities, cost more than a structure of wood.

In small houses, sudden changes of the external temperature are much sooner and more sensibly felt than in large structures, whether they are constructed of wood or iron, which arises from the fact that the smaller volume of air confined within becomes more rapidly heated, and hence the change is the sooner felt. Supposing the circumstance to be more strikingly sensible in the case of small iron houses, — then all that is necessary to counterbalance it, is just a little more attention to ventilation, during sudden changes of external temperature.

For large structures iron is incomparably superior to wood, and even for forcing-houses we would decidedly prefer the same material. The contraction and expansion of metallic hot-houses may be dreaded in the Southern States, if built on a very small scale, and badly managed; but in structures of moderate size, this evil will be found practically of little importance, unless they are badly constructed, and negligently managed.

The finest horticultural structures that have yet been erected in Europe are made of iron, and no houses of any importance are now being erected of wood, which proves its superiority over the latter material. The great conservatory, or Palm-house, at Kew, is wholly of iron, constructed under the auspices of the most scientific men in England. The Botanic Society's conservatory, in the Regent's Park, (already spoken of,) is made of iron. The fine plant-houses in the Glasnevin Botanic Garden, near Dublin, are constructed of iron, and the quite unequalled range of forcing-houses at Frogmore, in Windsor Park, are also of iron. In fact, the most extensive horticultural erections in Europe are made of iron, and many others, now in course of erection, are being made of the same material.

Admitting that properly constructed iron houses would cost, at the outset, somewhat more than wooden ones, their lightness and elegance render them much superior in point of appearance and, when their durability is taken into consideration, they will undoubtedly, be found cheaper in the end. But the cost of construction will vary, according as the details are understood by the constructors; for if Mr. Resorr can make a vinery of iron as cheaply as of wood, then other tradesmen, when they have properly understood the nature of the work, will surely be able to do the same. The Palm-house at Kew was constructed by a tradesman from Dublin, while some of the most extensive hot-house builders in England lived within the sound of their hammers, and the material and workmen were all brought across the channel, costing nearly as much as if brought to America; yet the workmanship was superior, and the cost said to be less, — proving that practice and knowledge of the details lessen the original cost of construction.*

* As instances of comparatively easy transportability of iron hot-houses, we might mention, that the whole of the materials of the immense structure at Kew were manufactured and fitted together at Dublin, and transported from thence to London. The unequalled range of forcing-houses at Windsor, one thousand feet in length, was made at Birmingham, and fitted together in the works, before they were transported to their final destination. Now it would have been just as easy, and perhaps little more expensive, to have shipped them to New York, or Boston, or Philadelphia, or Baltimore. When this is done in England, how long will American enterprise be behind them? We prophesy, not long.

SECTION VI.

GLASS.

1. EXPERIMENTS which have hitherto been made, in regard to the physical properties of glass as a transparent medium, have been conducted, generally, on purely chemical principles, and mostly without reference to observed facts, as regards the growth of plants, excepting, perhaps, those of the most common and obvious character. Partly for this reason, and partly from careless negligence, hot-houses have long been, and still continue to be, glazed with material of a very inferior description. If any one doubts this, let him look at some of the finest hot-houses in the country, and he will easily perceive the truth of this statement; the sickly and scorched appearance of the plants under its influence, being far more painful than agreeable to the eye of any one who takes an interest in the vegetable kingdom. This evil, alone, renders the very best cultivation of no avail.

The most elaborate and practically useful investigations that have yet been made, in this department, are those lately undertaken, with the view of securing the very best material that science and art could produce, for the glazing of the great Palm-house at Kew. We cannot do better than present our readers with the following extract from Mr. Hunt's report to the committee, which we take from Silliman's Journal of Science and Art, vol. iv., p. 431.

"It has been found that plants growing in stove-houses, often suffer from the scorching influence of the solar rays, and great expense is frequently incurred, in fixing blinds, to cut off this destructive calorific influence. From the enormous size of the new Palm-house, at Kew, it would be almost impracticable to adopt any system of shades that would be effective, this building being 363 feet in length, 100 feet wide, and 63 feet high. It was, therefore, thought desirable to ascertain if it would be possible to cut off these scorching rays by the use of a tinted glass, which should not be objectionable in its appearance, and the question was, at the recommendation of Sir William Hooker and Dr. Lindley, submitted, by the commissioners of woods, &c., to Mr. Hunt. The object was to select a glass which should not permit those heat rays, which are most active in scorching the leaves of plants, to permeate it. By a series of experiments, made with the colored juices of the palms themselves, it was ascertained that the rays which destroyed their color belonged to a class situated at the end of the prismatic spectrum, which exhibited the utmost calorific power, and just beyond the limits of the visible red ray. A great number of specimens of glass, variously manufactured, were submitted to examination, and it was at length ascertained, that glass tinted green appeared most likely to effect the object desired, most readily. Some of the green glasses that were examined, obstructed nearly all the heat rays ; but this was not desired, and, from their dark color, these were objectionable, as stopping the passage of a considerable quantity of light, which was essential to the healthy growth of the plants. Many specimens were manufactured purposely for the experiments, by Messrs. Chance, of Birmingham, according to given directions; and it is mainly due to the interest taken by these gentlemen, that the desideratum has been arrived at.

" Every sample of glass was submitted to three distinct sets of experiments.

" First. — To ascertain, by measuring off the colored rays of the spectrum, its transparency to luminous influence.

" Second. — To ascertain the amount of obstruction offered to the passage of the chemical rays.

" Third. — To measure the amount of heat radiation which permeated each specimen.

" The chemical changes were tried upon chloride of silver, and on papers, stained with the green coloring matter of the leaves of the palms themselves. The calorific influence was ascertained by a method employed by Sir John Herschel, in his experiments on solar radiation. Tissue paper was smoked on one side by holding it over a smoky flame, and then, while the

spectrum was thrown upon it, the other surface was washed
with strong sulphuric ether. By the evaporation of the ether,
the points of calorific action were most easily obtained, as these
dried off in well defined circles, long before the other parts presented any appearance of dryness. By these means it is not
difficult, with ease, to ascertain exactly the conditions of the
glass, as to its transparency to light, heat, and chemical agency,
(actinism.)

"The glass thus chosen is of a very pale yellow green color,
the color being given by oxide of copper, and is so transparent
that scarcely any light is intercepted. In examining the spectral rays through it, it is found that the yellow is slightly diminished in intensity, and that the extent of the red ray is diminished
in a small degree, the lower edge of the ordinary red ray being
cut off by it. It does not appear to act in any way upon the
chemical principle, as spectral impressions, obtained upon chloride of silver, are the same in extent and character as those
procured by the action of the rays which have passed ordinary
white glass. This glass has, however, a very remarkable action
upon the non-luminous heat rays, the least refrangible calorific rays. It prevents the permeation of all that class of heat
rays which exists below, and in the point fixed by Sir William
Herschel, Sir H. Englefield, and Sir J. Herschel, as the point
of maximum calorific action, and it is to this class of rays that
the scorching influence is due. There is every reason to conclude that the use of this glass will be effectual in preserving
the plants, and at the same time that it is unobjectionable in point
of color, and transparent to that principle which is necessary for
the development of those parts of the plant which depend upon
external chemical excitation, it is only partially so to the heat
rays, and it is opaque to those only that are injurious. The
absence of the oxide of manganese, commonly employed in all
sheet glass, is insisted on, it having been found that glass, into
the composition of which manganese enters, will, after exposure
for some time to intense sun-light, assume a pink hue, and any
tint of this character would completely destroy the peculiar
properties for which this glass is chosen. Melloni, in his investigations on radiant heat, discovered that a peculiar green

glass manufactured in Italy, obstructed nearly all the calorific rays. We may, therefore, conclude that the glass chosen is of a similar character to that employed by the Italian philosopher. The tint of color is not very different from that of the old crown glass, and many practical men state, that they find their plants flourish better under this kind of glass, than under the white sheet glass, which is now so commonly employed."

We understand the glass employed in the Kew Palm-house has fully answered the intended purpose, viz., of obstructing the most injurious portion of the heat rays; and we have learned, also, that it has answered all expectations as to its influence on the health of the plants, although its perfect utility, in this respect, has been doubted by some practical men. We think, however, that an absolute decision on its merits, in this respect, is rather premature, as we should prefer seeing the plants attain a greater size, so as to fill the structure more completely, and their foliage reach nearer to the glass, before pronouncing definitely upon the calorific effects of the latter.

As to the appearance of this glass, it is altogether a matter of taste, which we consider ourselves having no right to question; and, upon the whole, we think it in this respect unobjectionable. When viewed obliquely, from a distance, it is slightly green, but when viewed from within, and at right angles to its surface, it is clear and nearly white. This kind of glass is highly worthy of the attention of glass-makers and horticulturists in this country, and we have no doubt, when its qualities have been fairly tested and made known, it will be extensively employed in horticultural buildings.

No kind of economy is more sure to defeat its end than using cheap glass in horticultural structures. Many suppose, if a house is merely covered with glass and made transparent, that all is well. We know this to be a common opinion; yet we are fully prepared to prove its falsity, not by mere assertion, but by indubitable facts, — facts so clear that the most ignorant in these matters will be convinced, from his own observation, and on a scale so extensive, as to justify the conclusions that have been drawn from them.

We know of nothing connected with the erection of horticul-

tural buildings so vexatious as having the roof glazed with
bad glass; plants of almost every kind are certain to suffer
under it. Knotted and wavy glass is the worst of all, as the
knots and waves form lenses, and concentrate the sun's rays
upon the plants, and that part on which the concentrated ray
falls is sure to be burnt. It cannot for one moment be doubted
that the glass used in the majority of horticultural buildings is
not only inferior, but is of the very worst description; and, on a
recent examination of one hundred houses, we found scarcely
one free from the defects here spoken of. Indeed, we are fully
aware of the difficulty of procuring really good glass, at reasonable prices, for glazing hot-houses. But there cannot be a doubt
that the money saved is money lost; and if the vexation and
annoyance subsequently incurred by the use of inferior glass, be
taken into consideration, few persons of sound judgment will
hesitate in paying an increased price.

No doubt many of our readers will suppose that we are
unnecessarily particular on this point, but our experience has
taught us a severe lesson, and one, too, which no doubt has
been strongly impressed upon the mind of every gardener, of
lengthened experience in these matters. Against such an evil
there is but one resource, — and a bad one it is, — which is
shading, either by means of cloth blinds, or by painting, the
worst method of the two; but the one or the other is absolutely
necessary. The first is troublesome, the other is unsightly;
and, to be done right, both are expensive. We have a large
house now under our management, on which the glass is so bad
as to render its opacity absolutely necessary to prevent burning,
even when the sun's rays have lost their meridian power.

In very small houses bad glass may be used with less chance
of injury, as they may be easily shaded with blinds during the
noonday sun; but in very large structures this is only accomplished at very great expense; and in curvilinear houses, and
houses with irregular roofs, covering them with blinds is almost
impossible. Painting the glass, then, is the only resource,
unless glass be used which does not require it.

Little has been said on the effects of glass used in hot-houses,
by writers on practical horticulture. Although facts are obvious

and familiar in regard to it, yet the evils seem to be passed over as results which cannot be prevented. We can at this moment point to houses standing side by side, in one of which it is impossible to grow, and keep in health, any species of vegetation whatever, — no matter how hardy the tissue of the foliage may be, — without shading the glass almost to opacity; while, in the other, plants with tender and delicate foliage stand comparatively uninjured. The cause is obvious: the glass with which the one is glazed is full of waves and blotches, and altogether of the worst description; while that of the other, though not the best, is yet of better quality. The poorer glass burns vegetation, even when the incidental angle, between the impinging ray and a perpendicular to the roof, is as much as 45°.

From what has been already said regarding the influence of the different solar rays on vegetation, and, more especially, the experiments made with regard to the Palm-house at Kew Gardens, by which it has been found possible to manufacture glass which is opaque to the scorching rays, without at the same time obstructing the light, heat, and chemical rays which are essential to the development of plants, there can be no doubt that the scorching of vegetation in hot-houses, which has long been a serious drawback in exotic horticulture, can be prevented. And when more extended experiments have been made, a good material for glazing can undoubtedly be manufactured at a price that will insure its universal adoption in horticultural structures. It is to be earnestly desired that some of our enterprising manufacturers, — a class so remarkable for their fertility of invention, — will take up the matter seriously, and supply us with the material which exotic horticulture so much requires.

2. *Glazing.* — Common sash-glazing is generally performed with a lap of from one to three fourths of an inch, and, by many, with a full inch lap. This is a most objectionable method, as the broader the lap the greater the quantity of water retained in it by capillary attraction, and, consequently, the greater the breakage of the glass; for when the internal temperature falls, and this water becomes frozen, the glass is certain to crack in the direction of the bars. The lap should never be broader than

10*

a quarter of an inch, but where the panes or pieces of glass are not above five inches wide, one eighth of an inch is sufficient. Half an inch in roof-sashes, unless they are placed at an angle of not less than 45°, is almost sure to produce breakage, excepting the temperature within be kept sufficiently high to prevent the water retained between the panes from freezing.

Broad laps are objectionable, also, on other accounts; for the broader the lap the sooner it fills with earthy matter, forming an opaque space, and these spaces are so numerous as to have a very considerable effect upon the transparency of the roof, which is injurious by excluding the light, and is also unsightly in appearance. It may be puttied, but its opacity is the same, and its appearance no better than if filled with dirt. Where the lap is not more than one fourth of an inch, it may be puttied without any very disagreeable effect, but if the glass be perfectly smooth in the edges, puttying is useless, and the glass is better without it.

The most approved practice as to the laps, whether in roofs or common sashes, is, to make the breadth of the lap equal to the thickness of the glass, leaving it entirely without putty. But it is extremely difficult to get glaziers to attend to this, and it can only be obtained by employing good workmen, and keeping strict supervision over the work. This is not only the most elegant of all modes of glazing, but the safest for the glass, which, as we have observed, is seldom broken by any other natural means but the expansion of frozen water retained between the laps. This mode is also by far the easiest to repair, and is more durable than any method of filling the laps with putty, or with lead.

There are various other modes of glazing, as the lead and copper-lap methods, which, however, are so very objectionable as to be unworthy of occupying space in our description. The methods of shield glazing are equally objectionable, and little used. Curvilinear glazing has been used somewhat extensively and is, in the opinion of some men of undoubted skill, superior to the other methods already spoken of.

Curvilinear lap-glazing appears preferable to the square mode, for various reasons, one of which is, that the curve has a tendency to conduct the water to the centre of the pane, which is let out by a small opening at the apex of it. If the lap is broad, however, the water is accumulated by attraction precisely in the point where it is calculated to do most injury, — acting, in fact, as a power on the end of two levers of the second kind. But when the lap is not more than one sixteenth of an inch in width, no evil of this sort can happen.

It ought to be borne in mind that puttying, or otherwise filling up the laps, is in no case necessary if care be taken of the glazing, and smooth glass be used, and if the lap never exceeds one fourth, nor falls short one sixteenth, of an inch. However careful the laps may be puttied, in a very few years the putty begins to decay by absorption of moisture, and, when evaporation is great within, it becomes saturated with water, which readily freezes in frosty nights, (unless the temperature of the house is adequate to prevent it,) and breakage of glass is inevitable.

Reversed curvilinear glazing consists in making the lower edges of the panes to curve inwards, in a concave form, instead of curving outwards, in the common way. The effect of this method is the throwing of the condensed moisture down upon the bars, and thus conveying it off at the bottom of the roof, which prevents the moisture from being retained in globules, and dropping down upon the plants. This method is nothing more than reversing the position of the panes in common curvilinear glazing, and is, according to our opinion, preferable to it.

These are the most common and approved modes of glazing, although some others have been used that have not proved worthy of general adoption. Ridge-and-furrow roofs may be glazed in the same way. The size of the panes used makes no difference, — large ones only tending to reduce the opaque surface. Anomalous surfaces may be glazed with panes according to the figures of the bars.

3. *Color of Walls.* — The color usually applied to hot-houses is white. As affording the finest contrast with the plants in the interior, and the vegetation around the outside of the house, the general taste is manifestly in favor of this color; and, as it is

the best reflector of light, it is, also, on that account, preferable to any other. There are some considerations, however, in favor of a dark color, which, as has been already stated, absorbs a larger quantity of heat, and parts with it again on the cooling of the atmosphere. A yellow color we consider the most objectionable of all, both on account of its contrasting badly with the glass of the house and the verdure of vegetation, as well as the effects produced by it on the light, which, as will be seen from the preceding investigations, exercises an injurious influence on vegetation. The influence may not be so great in the reflected light, as when permeating yellow or orange-colored media, but the power is, nevertheless, exercised to some extent. The same investigations show the beneficial influence of a blue, or dark color, which perfectly accords with our observations on plants growing against dark bodies, otherwise exposed to abundance of light; and, when it is in accordance with the taste of the proprietor, we think the interior walls of hot-houses should be of a dark color.

In England, where the rays of light are less powerful than here, dark-colored walls are now very common. There, *light* is a more important consideration than heat : the latter can be applied by artificial means ; — not so the former. This probably tends to prevent the adoption of a dark color for the interior of their hot-houses. Here, dark walls are more desirable than white, as they absorb the heat-rays, during a powerful sun, and prevent the atmosphere from becoming so rapidly hot. This fact is sensibly felt on standing before walls of the different colors during the mid-day sun. By a white wall, the rays are reflected from the wall back into the air, or on any other body which is near it, by which the temperature of the air and the body is very much increased. A dark-colored wall, on the contrary, retains the heat which falls on its surface; and though it may *feel* colder, it contains more latent heat, which it only parts with when it is abstracted by the reduced temperature of the atmosphere. This, alone, is a good argument in favor of dark-colored walls in lean-to hot-houses.

The inner side of the rafters, astragals, and sash-bars, should approach to the color of the glass. As the light-rays do not fall on them, nothing is gained by making them dark, and it gives the house a heavy and gloomy effect. The structure is, or should be, transparent. The impression on the mind is that of a house covered with glass; and, as the rafters and astragals are only there as supports to the glass, they should be deprived, as much as possible, of their opaque character. When they are painted a dark color, the reverse effect is produced. A glaring white color is, also, objectionable; it is hurtful to the eye, and generally displeasing to a refined taste: some of the different shades of cream, or light stone color, will be more effective and pleasing. The same may be said in regard to the external portions of the roof. It may, by way of contrast, be a shade or two darker than the interior; but a decidedly dark color should be avoided. We have seen various plant-houses painted dark, and even dark red, but have seen very few who admired them. We do not wish to incur censure by finding fault with the taste of those who may fancy these colors, and admit that every one has an undoubted right to gratify his own taste. We give our opinions for the benefit of those who may choose to adopt them.

It is a good plan to give the wood-work of the structure a coat of some anti-corrosive paint before the color is put on. The timber is preserved much longer; and the house requires less painting, as the timber is hardened, and more impervious to moisture. For numerous preservative solutions, see Table **XVIII**, Appendix.

SECTION VII.

.. *Form of the Garden.* — The form of the garden must be determined by two conditions : first, the natural disposition of the ground chosen for its site ; and, secondly, by the aspect and position of the walls and hot-houses. If there are no hot-houses or walls, the form of the garden will be regulated mainly by the first condition. In most kitchen or culinary gardens, of any importance, if no walls are erected, wooden palings are generally substituted for them, which also regulate the disposition of the ground. The site having been fixed upon, with due regard to the considerations necessary in choosing the site for horticultural structures, (see Sect. I,) these considerations being in both cases equally applicable, the next thing to be done is the disposition and formation of the walks, which also define the size and shape of the borders and principal compartments of the garden.

2. *Walks.* — The principal walks from the house to the garden should be somewhat broader than the garden walks, and should, if possible, enter the garden at the south side. This is more especially desirable if there be hot-houses on the south side. In either case, however, it is desirable, as a more favorable impression is produced on the mind of the spectator than if entering at either side. The north side is the very worst for the principal entrance, as the necessary offices connected with the garden, — the mould-heaps, rubbish-piles, manure, &c., — are generally located in that quarter; besides, the impression, produced by the best trained trees on the walls or fences, and the general view of the ground, is lost. Next to the south, the east or west sides should be chosen.

There are various methods of forming walks, according to the character of the soil and sub-soil, and the kind of material at hand to form a surface. Where the ground is naturally wet, or where there is a liability of the accumulation of water, the soil should be taken out to the depth of at least twenty inches, — the section formed by the excavation forming an obtuse angle towards the centre, or forming the segment of a circle. These excavations should lead into drains, at the lowest points, to carry off the water that percolates through among the stones with which they are filled. They may be filled to within two inches of the intended surface of the walk, — the largest in the bottom, and the smaller toward the surface. This forms a durable and dry walk at all seasons; and, where the soil contains a considerable quantity of stones, which have been thrown out in the process of trenching, or the rubbish of building-materials, this affords a good medium of getting them out of the way.

On dry, gravelly ground, however, these excavations are useless, so far as drainage is concerned; and, shovelling aside the mere surface-soil, the walk may be laid down on the substratum beneath it. If the walks are on a level, or nearly so, the water generally finds its way off as quickly as it falls, and the cost of excavation is saved.

The surface of walks may be formed of grass, gravel, or sand. Good gravel is the best, sand the very worst, and grass can only be introduced with propriety in particular places. Sand, or loose gravel, makes a very uncomfortable walk, and, when of great length, is tiresome and disagreeable to walk upon.

A very common error, among those not acquainted with the proper method of making walks, is, to lay on too much surface-material; and, in many places, we have seen trenches taken out for walks, and filled, to the depth of a foot or more, with gravel, which, if laid on a hard surface to the depth of an inch or less, would have made a good walk, but which, at such a depth, all the walking, rolling, and pressing of years could never make it bind. It requires more skill than is generally supposed to make good walks. Among all the operations of the garden-maker there is scarcely one which we are so much disposed to find fault with, as in the making of walks; and this is precisely our reason for adverting to a matter which is apparently irrelevant to the general character of the present work.

The durability and comfort of walks consist chiefly in their power of resisting the action of the feet in walking on them, at all seasons of the year. Soft gravel walks, that yield no resistance to the motion of the body, are obviously unfit for being in a place where frequent walking is resorted to. Sand, also, makes a pretty walk to look at, but should never be employed where a good hard walk is required, unless it naturally possesses the property of binding.

It is quite possible, however, to have a hard solid walk, capable of resisting the action of the feet, and yet appear to have a gravelly or sandy surface, which is frequently admired. This is effected by preparing the lower strata of open material, then a substratum of binding material, and lastly, a thin layer of whatever material is wished for the surface, which should be sifted before being laid on. It is then well watered, if dry; then rolled well in, which has the effect of mixing it with the binding stratum beneath, and leaving a smooth surface, that becomes harder the longer it is used. In making up the substrata, it is necessary to tread each layer firmly as it is made up, so that no hollows or inequalities may occur on its subsidation, and subsequent use.

It must be remembered that the material of which the surface of a walk is composed, will not bind by any mechanical means, unless it contains something of a binding nature within itself. Clean gravel will not bind by any degree of mechanical pressure, unless it contains something to induce a general compactness and solidity over the whole surface.

The best material which we have met with in this country, and which is no doubt abundant in many places, is a kind of soft decomposing sandstone rock, containing a large quantity of oxide of iron. It must be laid down where it is finally to remain, when newly taken out of the pit, then subjected to a good shower of rain, or watered, and afterwards rolled or well trodden with the feet; it makes a solid walk, nearly as compact as the rock itself. It may be objectionable on account of its color, but this is easily changed by a thin layer of any other material on its surface, which partly mixes and binds with it.

Broken, or what is sometimes called rotten rock, containing oxide of iron, is to be preferred to gravel for making a surface, and pit gravel is to be preferred to river or sea gravel, as it contains generally more oxide and earthy matter. Clay forms a good under-surface, and when thinly covered with small gravel and well rolled, forms a most excellent and durable walk. Common gravel may also be mixed with coal ashes and lime rubbish, which tends to bind it, and also with common garden soil; but this is a last resource. Gravel mixed with earth, and more especially vegetable earths, has a great tendency to produce weeds, and is therefore very troublesome to keep clean. It also readily absorbs moisture and becomes soft in wet weather. and especially during winter frosts.

Where expense is not spared, a composition may be made, consisting of small shell gravel, or pounded granite, about one tenth part of brick-dust, and cement, mixed together. This, laid down upon a firm, prepared surface, in a wet state, and well rolled, will form a surface as hard as marble.

The form of the surface should be nearly flat; grass walks should be completely so; gravel walks may rise slightly towards the middle, but not so much as to affect the convenience of as many persons walking abreast as the breadth of the walk will admit. A walk six or eight feet wide should not fall more than an inch towards each side, this being sufficient to throw the water that falls on it towards each side, without being any inconvenience to pedestrians occupying its whole breadth.

If the walk be edged with turf, the crown of the walk should, when finished, be on a level with the turf at each side, one inch being quite enough for depth of edging; besides, the walk generally subsides, while the verges become higher, for which an allowance must be made. The same rule applies to walks edged with box, which are most suitable in a kitchen garden.

3. *Borders and Interior Compartments.* — The width of the borders and size of the compartments must be regulated by the height of the wall or fence, and the extent of the garden. The

11

best general rule that can be laid down is to make the breadth
of the borders equal to the height of the wall or boundary fence,
whatever it may be; they may be made broader, but not nar-
rower, for then they produce a bad effect; a narrow border
beside a high fence is very displeasing to the eye.

The size and number of the compartments are determined by
the number and disposition of the walks. It is decidedly a bad
plan to have too many walks, as the ground is not only taken
up with them, which require a deal of labor to keep them clean,
but the effect of the garden is lessened. If less than two acres
be enclosed, a walk running parallel with the boundary, say
twelve feet distant from it, and another intersecting the garden
in the middle, running south and north, will be sufficient; if
more than two acres be enclosed, another intersecting walk, running east and west, may be introduced. If the garden be
worked by horse labor, the larger the compartments the better;
if wrought entirely by manual labor, these compartments may
be sub-divided for the crops, by rows of fruit-trees, or fruit-
bushes, as may be required. It should be observed, that to
have a few walks, and those of good width, gives the garden a
better appearance, and is in every way preferable to having a
large number of contracted ones, and it leaves the compartments
to be sub-divided by alleys or other means, as may be most convenient for access to the crops.

In many gardens, trellises or espalier rails are adopted. The
proper place for an espalier rail trellis is on the inside of the
principal walks, leaving a border of at least six feet. Many
gardeners condemn them, and perhaps justly, in small gardens,
as it confines the ground too much; but in large gardens, espa-
liers, if well managed, are both useful and ornamental. The
railing should be plain and neat, not more than five or six feet
high, with the upright rails, to which the trees are tied, about
eight inches apart.

It is not our purpose, at present, to dwell on the laying out
of gardens. We have merely adverted to the subject, in so far
as it is connected with the object of this treatise.

Walls. — As garden walls may be regarded as horticultural structures, we will here make a few remarks upon them.

In Europe, walls are built around gardens of all kinds, whether the enclosed space be one or twenty acres. Their chief use is for training the more tender kinds of fruit-trees upon their southern aspect. The enclosed space is generally appropriated to the growth of culinary vegetables, and containing also the hot-houses, which occupy a part of their south aspect. These gardens are of various forms, and we have seen them circular, oval, square, and oblong. The latter shape, with the angular corners cut off, is undoubtedly the most desirable shape for a vegetable garden. The oval and polygonal forms are preferred by some, on account of their affording a more equal distribution of sun and shade. · But we are at a loss to find out how this can be the case, as, however a wall may be placed, it can only obtain a certain amount of direct sunshine during the day, and the inconvenience resulting from the adoption of these forms is very considerable, both in the management and culture of the interior compartments, and in the training of the trees. Moreover, an equal distribution of sunshine is not so desirable as may appear; as, while the warmest portion of the wall may be appropriated to the more delicate and early fruits, the coldest, or northern portion, may be as profitably appropriated to late sorts, or for retarding earlier kinds, both of which purposes are as useful as an early aspect.

In this country, walls have been little employed in the formation of gardens, and only in a few places have they been adopted, as at the fine gardens of Mr. Cushing, at Watertown, and Col. Perkins, at Brookline, in the vicinity of Boston, — two of the finest gardens in this country. Some other places have also portions of walls surrounding the garden, but we have seen none where any principles of design have been adopted and carried out so much as at the former place.*

* In Hovey's Magazine of Horticulture, pp. 50—53, vol. xvi., we have described the beautiful gardens at this place, from a visit which we gave them at that time. We have subsequently visited them, as well as many other places, and still consider them the finest gardens we have seen in America. They are made precisely in the style of modern

In nearly all gardens, trellises and wood fences are employed instead of walls, as enclosures to the garden ground; and these are well adapted for the purpose, as the fruits which require the protection of walls in England thrive and produce their fruit in greater perfection as open standards here.   The utility of walls, however, around a garden, cannot be doubted, even in this country, especially as regards the protection they afford to trees trained on them, in early spring.   Walls may be considered as useful to plants trained on them, or near to them, in three ways: — first, by the mechanical shelter they afford against cold winds; secondly, by giving out the heat they had acquired during the day; and, thirdly, by preventing the loss of heat which the trees would sustain by radiation.   [See Experiments by Dr. Wells, in the third part of this work, Section VI. *Protection of Plant-houses during Night.*]

The same arguments which have been applied in favor of the best aspect for hot-houses, [see Section I.,] are equally applicable to walls.   In the middle and southern states, we should think walls having a due southern aspect decidedly objectionable, and, for tender and delicate kinds of fruit-trees, would decidedly prefer either a south-eastern or a south-western aspect.

The height of walls, or fences of any kind, round a garden, should always correspond to the space inclosed.   Twelve feet may be taken as a maximum height.   In England, low walls produce a greater effect in accelerating fruit than high ones;

English gardens, surrounded with fine walls, with the principal range of hot-houses, about 300 feet in length, on the southern aspect of the wall on the north side of the garden, and a smaller range on the inside of the east and west walls, all lean-to houses.   There are convenient back-sheds and other offices on the north side of the hot-houses.   There is no wall on the south side of this garden, which we think is very appropriately dispensed with.   We regard this as a general rule, and more especially in gardens of small size, as it gives the enclosed spaces a less meagre and confined appearance.   This garden, alone, of any which we have seen in this country, bears an impress of the style and genius of Loudon.   And though we have some faults to find with the surrounding grounds, nevertheless, we believe, taking it all in all, it is the most perfect specimen of modern European gardening in this country.

but in this country the great radiation of heat from the earth, during the heat of summer, would render low walls of little use. On the other hand, high walls have always a gloomy effect, and, where it is necessary to have high walls round a garden, it is better to relieve the monotony of the wall by making it of different heights.

Hot, or flued, walls are very common in European gardens, and have been used upwards of a century; and, in our opinion, where walls can be of any importance in this country, in the practice of horticulture, it must be chiefly as flued walls. In summer, the protection of a wall is not required to ripen the common fruits, and in hot summers they are frequently injurious, by the attraction and radiation of heat during the midday sun, by which the leaves are sometimes scorched. It must be as protectors of peach and apricot blossoms in spring, and accelerating the ripening of grapes in autumn, in which they can be most serviceable to the horticulturist; and for these purposes hot walls are of great benefit. [See Wall Heating, Part II., Sec. V.]

Flued walls can be built as cheap, if not cheaper, than solid ones, and are invariably built of brick; indeed, a considerable saving of material is effected, as little more than half of the bricks required to build a solid wall will build a hollow or flued wall; and, unless a flued wall be desired, it is better to dispense with a wall altogether, for although a wooden paling will not absorb so much heat as a brick wall, as a structure for mechanical shelter it is in every way equal to it, providing it be boarded perfectly close, and sufficiently high. The comparative cheapness of wooden fences, for gardens, must give them the preference, and the comparative beauty of brick walls and wood palings is a matter of taste which must be decided by the proprietor.

Walls, or close palings, must, in all cases, be faced with a light trellis, made of laths or wire, to which the trees can be trained. The injury resulting to trees nailed on walls, in our gardens, is owing to their touching the material of the wall. The branches should be trained at least six or eight inches from the surface, so as to admit a stratum of air between the wall

11*

and the branches.   When this is attended to, no injury results
to the foliage, even in the hottest of seasons.

Boarded walls have long been used in northern countries, and
are frequently made to incline considerably towards the north,
so as to present a better angle to the sun's rays than if standing
upright ; an expedient which here is unnecessary.

We cannot help thinking that flued walls are worthy of more
attention from horticulturists than they seem to have had, especially when early fruit is desired, without the trouble and expense
of a glazed structure, as an expedient for a hot-house.   [See cut
50, in the next part of this work, page 245.]

PART II. HEATING.

SECTION I.

PRINCIPLES OF COMBUSTION.

1. To warm hot-houses, etc., most economically and efficiently, we must study not only the *principles of heating*, but, also, the *principles of combustion*. And as we are yet far from having obtained a complete knowledge of the most profitable manner of submitting coal and other kinds of fuel to the process of combustion, or, of applying the caloric so obtained to increase the temperature of hot-houses, it will, therefore, be desirable to begin at the beginning of this part of our work, and before treating on the different mechanical contrivances in common use for the generation and diffusion of heat by combustion, let us first consider the principles upon which these ends are to be obtained.

The subject before us involves a consideration of the nature and properties of the various kinds of fuel. It examines the chemical action of their several constituents on each other. It applies those inquiries to the class of chemical results which may be useful, and avoids those which are injurious. It involves also, in an especial degree, the closest observation on the separate influences which each of the constituents of atmospheric air exercises on combustible bodies, in the generation of those extraordinary elements of nature, *heat* and *light*. And, finally, it investigates the cause and character of *flame* and *smoke*, and the influence these have on the former.

Economy of fuel being one of the most important points to be sought for in a heating apparatus, we must inquire whether our common furnaces be so constructed as to give us the maximum quantity of caloric, for the fuel that is consumed. We, therefore, must look into the furnace, and consider chemically as well as practically, the operations which are there going on, so that we may improve its arrangements, and adapt them so as to give full practical effect to the several processes which constitute combustion.

To enable our practical readers to obtain a more accurate knowledge of the processes going on in the furnace, and of the results of the common mode of managing the fires of extensive forcing houses, we will enter more fully upon the constituents of coal, and the gases thereby generated, which form such an important part of the fuel itself, and which, by their escape into the atmosphere from the chimney, or into the atmosphere of the house from the flue, become the source of immense loss of heat. And, in the latter case, the loss is more than doubled, as they are destructive in the highest degree to every kind of vegetable life.

In undertaking to show how these evils may be remedied, we must not be understood to concur in the exploded opinion, that these gases may be consumed by the methods hitherto used for that purpose, viz., by passing the smoke over a body of red-hot fuel at a distance from the burning and smoking mass. And however desirable it may be to know of some way of preventing smoke from being emitted in clouds from the chimney of hot-houses, yet, if we can discover no other method of obviating the evil, except " burning it," according to the common acceptation of that word, I fear we must continue to put up with the loss and annoyance as it is.

It is not our purpose here to show how the smoke from fuel may be burned; but rather, we will attempt to show how fuel may be burned without smoke. And, let it be observed, this distinction involves the main question of economy of fuel.

When smoke is once produced in a furnace or flue, we believe it to be as difficult to burn it, (and convert it to heating purposes,) as to burn and convert the smoke issuing from the flame of a candle to the purposes of light. If, indeed, we could collect the smoke and unconsumed gases of a furnace, and separate them from the products of combustion which the flues carry off, they might, subsequently, be made instrumental to the purposes of heat; but, by the common method of constructing furnaces, their collection is impossible.

When we see smoke issuing from the flame of an ill-adjusted common lamp, the heat and light are diminished in quantity. Do we attempt to burn that smoke? No; it would be impossible. Again, when we see a well-adjusted lamp burn without producing any smoke, the flame is clear and white. But here, the lamp has not burned its smoke; it has burned *without* smoke; and it remains to be shown why the same methods may not be employed with regard to common furnaces, whereby they may burn without smoke, and thereby give out a greater quantity of heat, as in the case of the common and Argand lamp, since the elements of combustion in both cases are the same.

2. In pointing out the leading characteristics in the use of coals, it is unnecessary to enter into detail of the various processes of gasefaction. We will, however, give this part of our subject a little attention, as the greater portion of the practicable economy in the use of coal, and the management of furnaces, will be found more or less connected with the combustion of the gases which arise from the combustion of fuel, and as the numerous combinations of which they are susceptible embrace the whole range of temperature, from that of flame down to the refrigeratory point.

The subject of gaseous combinations, then, is undoubtedly an important part of our inquiry. And those who would study the economy of fuel, and the obtaining from it the greatest quantity of heat, cannot altogether dispense with the part of our subject which at present lies before us. Though it may not appear equally interesting and important to every one, it is, nevertheless, the alpha and omega of the whole process of combustion. The gardener may say, what has this to do with gardening? But we tell him, plainly, that this is an essential part of his business, which will be generally admitted by intelligent men, that so long as a furnace is connected with a hot-house, and fuel consumed in that furnace, this must necessarily be a part of his business.

On the application of heat to bituminous coal, the first result is its absorption by the coal, and the consequent disengagement of gas, from which all that subsequently bears the character of flame is exclusively derivable. This gas, whether it be in a close retort, or in a furnace, is associated with several other substances, more or less tending to deteriorate its inflammable properties and powers of giving out heat and light. In the preparation of gas, or smoke, for illuminating purposes, these impurities are separated, and the pure gas alone is used. As, however, this separation cannot be effected in a common furnace, and, as the entire gaseous products of the coal, good and bad, are indiscriminately consumed together as they are generated, it is the more incumbent on us to be cautious, lest, by any injudicious arrangement, we force these impurities into more active energy, and thus increase their deleterious power.

We will not stop here to consider the nature of those impurities arising out of the unions of sulphur, and the other injurious constituents of coal, although they exercise a mischievous influence on the calorific effect of the gas burning in the furnace, but will consider those constituents alone, which unite in forming the useful gases, and from which we are to derive heat.

These constituents are the hydrogen and the carbon. And the unions which alone concern us here, are, first, *carburetted hydrogen;* and, second, *bi-carburetted hydrogen*, commonly called olefiant gas. These two, and their unions with the air, in the process of combustion, we will shortly examine.

Gases, as well as other bodies, endowed with the power of giving out heat and light, have been called combustible. This term has been a source of much error in practice, from a misconception of its meaning, under the received impression that combustibles possess, in some undefined manner, and within themselves, the faculty of burning. And, though every person knows that they will *not burn without air*, still the part which air acts in the process is but little inquired into. It is but lately that the nature of this union of the gas with the air has come to be fully understood; and, although the abstract question as regards the immediate cause of that chemical action, which we call combustion, may continue to be disputed, and new theories continue to be broached, still, for all practical purposes, it is sufficiently defined and understood.

And here we are called on to inquire, with reference to the gases under consideration, whether there are any peculiar conditions which can influence the amount of heat to be obtained from them ? and, if so, what they are ? This, again, involves other questions in reference to *air*, and the part which it has to act in the process; and thus we find ourselves introduced into the *chemistry of combustion.*

One advantage of receiving the subject in this light, is, that we shall see how idle would be any calculations or arrangements as to the dimensions or details of a furnace, before we had well examined and understood the rationale of that process on which these details must necessarily be contingent. For what chemist would begin by deciding on the dimensions of his retort, or other apparatus, before he had considered the particular purposes to which they were to be applied ? Yet such is the every-day practice of those who profess to instruct us in these matters. The absurdity of this practice, and the dangers into which it leads practical men, will be more apparent when we come to consider the nature of heating apparatuses, and the powers and properties which belong to each.

Combustibility, then, is not a quality of the combustible taken by itself. It is merely a faculty which may be brought into action through the instrumentality of a corresponding faculty in some other body. It is, in the case now before us, the union of the combustible with oxygen, and which, for this reason, is called the *supporter*. Neither of which, however, when taken alone, can be consumed.

To effect combustion, then, we must have a combustible, and a supporter of combustion. Strictly speaking, combustion means *union;* but it means chemical union, — one of the accompanying incidents of this union being the emission of heat and light. What the nature of heat is, or how it is liberated during chemical action, it is not our province to consider; nor does it relate much to our present inquiry. Sufficient for our present purpose, is the fact, that the chemical union of the combustible, (the coal,) and the supporter of combustion, (the oxygen of the air,) is the cause of heat being given off; and, further, that exactly in the ratio that such union is complete, is the **quantity** of heat increased.

But we have not the means of obtaining this necessary supporter in sufficient quantity, in a separate state, except at an expense which would render it incompatible with the purposes of a furnace. Our only alternative then is to apply to the atmosphere, of which it forms a part, in order to satisfy our wants. Had we to purchase this oxygen, we would, necessarily, be more economical of its use, and inquire more respecting its application. But, finding an abundant supply at hand, in the atmosphere, and obtaining it without expense, we are careless of its use, and unconscious of its value, and take no note of the large quantity of the noxious ingredients with which it is accompanied, or loss sustained, by diminishing the supply; and hence, many of the evils, such as bad apparatus, bad fuel, and bad furnaces, might be easily remedied, were the properties of these gases fully understood.

The unions we have now to consider are those which take place between the constituents of the coal and the atmospheric air, namely, the hydrogen and carbon of the former, and the oxygen of the latter. Dr. Ure calls the carbonaceous part of coal, " the main heat-giving constituent." In this he must be understood to include that portion of the carbon which forms one of the constituents of the gases alluded to, and, although, for the purposes of the *furnace*, so much value is set upon the solid part — *the coke* — we must not, on that account, undervalue the heat-giving properties of *the gas*. Indeed, the extent of those powers is strikingly brought before us, by the fact, that for every ton of bituminous coal no less than 10,000 cubic feet of gas are obtained.

When we consider the immense heating powers of such a mass of flame as would be produced by 10,000 feet of gas, we cannot resist the conclusion, that there must be something essentially **wrong** in the mode of bringing it into action *within a furnace*, as compared to its well known efficacy in *an argand burner*. That this is the fact, will appear manifest as we proceed. And one of our objects is to show how greater heat may be obtained by the combustion of the volatile products of the coal, than by allowing the whole body of gas to escape into the atmosphere.

Let us bear in mind, that smoke is always the same, whether it may be generated in a common fire-place, in a furnace, or in a retort; and that, strictly speaking, it is not inflammable, as by itself it can neither produce flame nor permit the continuance of flame in other bodies, as is proved from the fact that a lighted taper being introduced into a jar of coal gas, (or smoke,) is instantly extinguished.

How, then, is it to be consumed or prevented, and rendered available for the production of heat? The answer is, solely by effecting a chemical union, not with the *air* merely, as is the dangerous notion, but with the *oxygen* of the air, — the "*supporter*" of flame, the heat-giving constituent of the air, in *given quantities*, and at a given temperature.

This at once opens the main question, What are these *quantities*, and what is this *temperature?* and, are there any other conditions requisite for effecting the chemical union of the oxygen of the air with the inflammable gas, *to the best advantage?*

Effective combustion, for practical purposes, is, in truth, a question more as regards the *air* and the *gas;* and the former, as referable to our object, would appear better entitled to the term combustible than the latter, inasmuch as the heat is increased in proportion to the quantity of air we are enabled to use advantageously. Besides that, we have no control over the gas after having thrown the fuel on the furnace, but we *can* exercise a control over the air, as we shall show, in all the essentials of perfect combustion. It is this which has done so much for the perfection of the lamp, and may be rendered equally available for the furnace.

Now, although this control, and the management arising out of it, influences the question of perfect or imperfect combustion, and, therefore, affects that of economy, yet, strange to say, in an age when chemical science is so advanced, and in a matter so purely chemical, this is precisely what is attended to in practice. The *how*, the *when*, and the *where*, this controlling influence over the admission and the action of the air is to be exercised,

12

are points demanding the most attentive consideration from all
who are interested in these matters.

Much confusion at present prevails in all that regards hot-
house furnaces, as well in their practical working as regards the
admission of air and the combustion of fuel. In commenting
briefly upon the constituents of coal smoke, or coal gas, carburetted hydrogen, and the quantity of air required for their
combustion, we will be as explicit as possible, without going
more into scientific detail than is consistent with the means and
opportunities of that class of practical men for whom we write.

3. The first step towards effecting the perfect combustion of any
combustible gas, is the ascertaining the quantity of oxygen with
which it will chemically combine, and the quantity of *air* required for supplying such quantity of *oxygen*. Here, then, we
are called on for strict chemical proofs — these several quantities
depending, not on the dictum of any chemist, but on the faculty
which each particular gas possesses of combining with certain
definite proportions of the other — the supporter; these respective proportions being termed "*equivalents*," or combining volumes. This doctrine of equivalents must, therefore, be understood before we can be prepared to admit the necessity of any
precise quantities. This question, as to quantity, is also the
more important when we consider that the quantity of effective
heat obtained by the combustion of any body, will be in exact
relation to the quantity of oxygen with which it will chemically
combine.

Let us begin, then, by inquiring into the constitution of the
coal gas, and the relative proportions in which its constituent
elements are combined, as these necessarily govern the proportions in which it will combine with the oxygen of the air.

Now, the doctrine of "equivalents," that all-convincing proof
of the truths of chemistry, being clearly defined and understood,
reduces, to a mere matter of calculation, that which would
otherwise be a complicated tissue of uncertainties. And let no
mechanic feel alarmed at this introduction to "elementary atoms"
and "chemical equivalents," or imagine it will demand a deeper
knowledge of chemistry than is compatible with his sources of

information; neither let him suppose he can dispense with the knowledge of this branch of the subject, if he has anything to do with the combustion of coal. Without it, he is at the mercy of every speculative " smoke-burning " pretender; whereas, with it, his mind will be at once opened to the simplicity and efficiency — I may add, to the truth and beauty, of nature's processes, as regard combustion.

There is not, indeed, a more curious or instructive part of the inquiry than that respecting the conditions and proportions in which the compound gases enter into union with the constituents of the air; neither is there one more intimately connected with the practical details of our furnaces. These introductory remarks are, therefore, necessary for those who are not already familiar with it. Indeed, without some information on this head, the unions of the gases might appear capricious or uncertain; whereas, in fact, they are regulated by the most exact laws, and subject to the most unerring calculations.*

---

* Mr. Parkes observes: — " We are unfurnished with any definite, determinate experiments regarding the proportions in which *air* and *fuel* unite during combustion. We are, practically speaking, altogether ignorant of the mutual relations which subsist between the *combustible* and the *supporter of combustion*, (the fuel and the oxygen ;) and, though we know that, without oxygen, we cannot elicit heat from coal, we have yet to discover the most productive combinations of the two elements.

" Here, then, remains a wide field for research and experiment, worthy, and, indeed, requiring the labors of a profound chemist."

These matters are now better understood, and those " *most productive combinations* " rendered familiar and certain, by the labors of that " profound chemist," John Dalton, who first drew the attention of the chemical world to the subject of equivalent proportions, and taught us the importance and necessity of ascertaining those proportions — in fact, of " *reasoning by the aid of the balance.*"

Dalton's papers were first read before the Manchester Philosophical Society, and published in their memoirs, in the year 1803. These volumes are very scarce, and I have not been able, anywhere, to meet with a complete copy of them. The Royal Institution, where Davy brought his great discoveries to light, contains but the five volumes of the first series. These volumes, or, at least, the papers of Dalton, should be republished, for the purpose of showing the correct chain of reasoning by which the mind of that acute philosopher proceeded.

Much of the apparent complexity which exists on this head arises from the disproportion between the relative *volumes*, or *bulk*, of the constituent atoms of the several gases, as compared with their respective *weights*.

For instance, an atom of *hydrogen* (meaning the smallest ultimate division into which it is supposed to be resolvable) is *double* the bulk of an atom of *carbon vapor ;* yet the latter is *six times the weight* of the former.

Again, an atom of hydrogen is double the bulk of an atom of oxygen ; yet the latter is *eight* times the weight of the former.

So of the constituents of atmospheric air, nitrogen and oxygen. An atom of the former is double the bulk of an atom of the latter; yet, in weight, it is as fourteen to eight.

A further source of apparent complexity arises from the faculty of condensation, or diminution of bulk, which, in certain cases, attends the *union* of the gases. For example, one volume of oxygen and two volumes of hydrogen, *when united*, condense into a volume equal to that of the hydrogen alone, (the weight being, of course, the sum of both;) that is to say, one cubic foot of oxygen chemically combined with two cubic feet of hydrogen condense into the bulk of two cubic feet : and so on, each union bearing its now ratio of volume and weight. This apparent complexity, however, we shall soon see give way to a systematic consideration of the subject.

We have stated that there are two descriptions of hydro-carbon gases, in the combustion of which we are concerned; both being generated in the furnace, and even at the same time, namely, the *carburetted* and *bi-carburetted* hydrogen gases. For the sake of simplifying the explanation, I will confine myself to the first, as forming the largest proportion of the gas to be consumed, namely, the carburetted hydrogen, or common *coal gas*, as I shall call it for the sake of brevity.

Now as, during combustion, the atoms of this gas become *decomposed*, and its constituents *separated ;* and as these will be found to exercise separate influences during the process. it is essential that we examine them as to their respective properties weights, and volumes.

On analyzing this mixed gas we find it to consist of two volumes of hydrogen and one of carbon vapor; the gross bulk of these three being *condensed into the bulk of a single atom of hydrogen;* that is, into two fifths of their previous bulk, as shown in the annexed figures. Let figure A represent an atom of coal gas — carburetted hydrogen — with its constituents, carbon and hydrogen; the space enclosed by the lines representing the relative size or volume of each; and the numbers representing their respective weights — hydrogen being taken *as unity* both for volume and weight.*

Carburetted Hydrogen.   *Bi-carburetted Hydrogen.*

A.

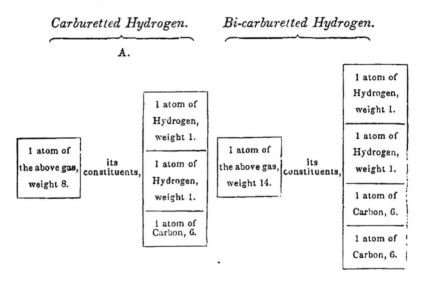

* "Ce gaz (carburetted hydrogen) est composé de 75.17 parties (by weight) de carbone, et 24.33 d'hydrogène; ou, d'un *volume* de carbone gazeux et quatre volumes de gaz hydrogène, condensés à la moitie due volume de ce dernier, ou, aux 2/5 du volume total du gaz, de manière que de cinq volumes simples, il n'en resulte pas plus de deux de la combinaison."— *Berzelius*, vol. i., p. 330.

12*

Or they may be represented thus:

Carburetted Hydrogen.

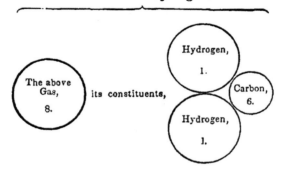

Bi-carburetted Hydrogen.

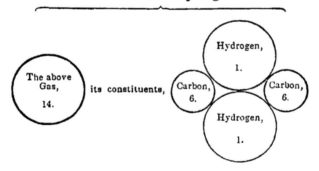

Although not intending to take any further notice, in this place, of the *bi-carburetted* hydrogen, I have, however, annexed the above diagrams, representing this gas and its constituents, that both may be under view at the same time ; and by which it will be seen, that although, *in volume, the two gases are precisely the same, there is yet double the quantity of carbon in the bi-carburetted that there is in the carburetted hydrogen :* this circumstance is of great importance, and must be kept in our recollection, as these proportions will be found to have a considerable influence during the subsequent process of its combustion. *

* The mode of representing the volumes of gas, by *rectangular* figures, as adopted by Mr. Brande and other chemists, is favorable, so far as *single atoms* are concerned, inasmuch as the eye at once recognizes the

I would here observe on the importance of keeping in mind this double relation of *weight* and *volume*, and the atomic constitution of these gases, as it will prevent much of that confusion which too often embarrasses those who are not familiar with the subject of gaseous combinations.

Let us now, in the same analytical manner, examine an atom of atmospheric air, the other ingredient in combustion. .

Atmospheric air is composed of two atoms of nitrogen and one atom of oxygen: and here again we find a great disproportion between the relative volumes of these constituents; one atom of nitrogen being *double* the volume of an atom of oxygen, while their relative weights are as 14 to 8: the gross *volume* of the nitrogen, in air, being thus four times that of the oxygen; and in *weight*, as 28 to 8, as shown in the annexed figure.

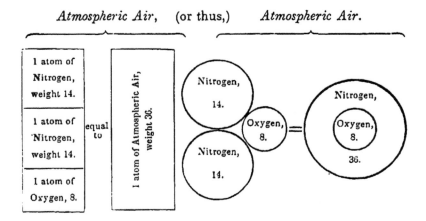

Here we are relieved from the complexity arising out of any difference in *volume* between these constituents, when *united* and when *separate*. In the coal gas we found the constituents condensed into *two fifths* of their gross bulk when separate: this, we see, is not the case with *air*; an atom of which is the same, *both as to bulk and weight*, as the sum of its constituents.

relation between *volumes* and *half volumes*. As, however, I shall have to do with *masses* of these gases, I have adopted *circular* figures, the relation between the sizes of the volumes of the different gases being the same.

Thus, we find, the oxygen — the *heat-giving* constituent **of**
the air — bears a proportion in volume to that of the nitrogen, as
1 to 5; there being, in fact, but 20 per cent. of oxygen in atmos
pheric air, and no less than 80 per cent. of nitrogen; a circum
stance which should never be lost sight of in all that has to do
with its admission and application.

Having shown the composition of coal gas, and also of air
with the weights and volumes of their respective constituents,
we now proceed to the ascertaining the *separate quantity of oxy
gen required by each of those constituents*, so as to effect its per
fect combustion, and produce the largest quantity of available
heat; in other words, to find the "*chemical equivalent*," or vol
ume of air, required for the saturation of this *mixed gas*.

Now, this is to be decided, not by the quantity of air we may
admit or force into the furnace, but solely by the faculty with
which each of these constituents is endowed of uniting *chemically*
with the oxygen.

With respect to this power, or faculty of reciprocal saturation,
the first great natural law is, that *bodies combine in certain fixed
proportions only*, — a remarkable feature in this law, as far as
gaseous bodies are concerned, being, that it has reference both to
volume and *weight;* thus, by their concurrence, establishing the
principle which now no longer admits of any doubt. *

The important bearings of this great elementary principle of
proportionate combination cannot be more strikingly illustrated,
or its influence rendered more familiar, than in the several com-

* " L'experience a demontré que, de meme que les élémens se combinent dans des proportions fixes et multiples, relativement a leur *poids*,
ils se combinent aussi, d'une manière analogue, relativement à leur
volume, lorsqu'ils sont à l'etat de gaz : en sorte qu'un volume d'un
élément se combine, ou, avec un volume egal au sien, ou avec 2, 3, 4 et
plus de fois son volume d'un autre element à l'etat de gaz. En com-
parant ensemble les phénomènes connus des combinaisons de substances
gazeuses, nous découvrons les *memes lois* des proportions fixes, que celles
que vous venons de deduire de leurs proportions *en poids :* ce qui donne
lieu à une manière de se representer les corps, qui doivent se combiner,
sous des *volumes* relatifs à l'etat de gaz. Les degres de combinaisons
sont absolument les mémes, et ce qui dans l'une est nommé *atome*, est
dans l'autre apellé *volume*." — *Berzelius*, vol. iv., p. 549.

binations of which the elements of atmospheric air are susceptible, and the extraordinary changes of character and properties
which accompany the changes, in the relative *quantities* alone,
of the combining elements.

For instance, oxygen unites chemically with nitrogen in five
different proportions, forming five distinct bodies, each essentially
different from the others, thus:

Atoms.	Weight.		Atoms.	Weight.		Gross Weight.
1 of Nitrogen	14 unites with	1 of Oxygen	8 forming	Nitrous Oxide . . 22		
1 "	14 "	" 2	"	16	"	Nitric Oxide . . 30
1 "	14 "	" 3	"	24	"	Hyponitrous Acid 38
1 "	14 "	" 4	"	32	"	Nitrous Acid . . 46
1 "	14 "	" 5	"	40	"	Nitric Acid . . . 54

Or thus:

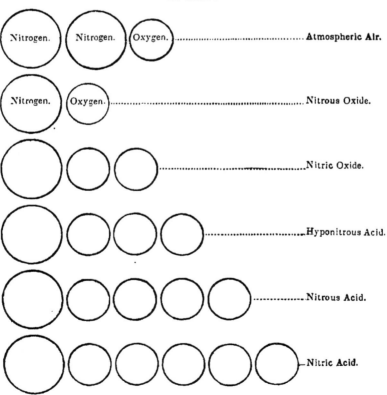

A description of the properties of these distinct bodies may be found in any chemical work of authority, and I only mention these unions to exemplify the importance of attending to the *proportions* in which bodies unite; as we here find the very elements of the air we breathe, by a mere change in the *proportions* in which they are united, forming so many distinct substances, from the *laughing gas*, nitrous oxide, up to that most powerful and destructive agent, *nitric acid*, commonly called *aqua-fortis.*

This case of the combination of nitrogen and oxygen also shows the·importance of the distinction between *mechanical* and *chemical* union; these two elements being only *mechanically* united in forming *atmospheric air*, by which the essential properties of its two constituents as preserved unaltered; whereas, in the five bodies above enumerated, the union is *chemical*, and, consequently, the essential characters of their respective constituents are lost, and new ones obtained.

Now, to apply these principles to the bodies under consideration, namely, the *carbon* and *hydrogen*, and ascertain the proportions of oxygen they respectively require to produce chemical union.

These two constituents, though united in the one body — the gas — yet, not only separate themselves during combustion in a remarkable manner, but, *by two distinct processes, form two essentially different unions.* This is an important feature of the development of chemical action which the law of equivalents at once points out and enables us to satisfy, although this *double process* does not appear to be understood, much less to be provided for, *in practice*, though familiar to every chemist.

On the first application of heat, or what may properly be termed the firing or lighting the gas, when duly mixed with air, the carbon *separates itself from its fellow-constituent, the hydrogen*, and forms a union with the former, the produce of which is *carbonic acid gas.*

Now, the laws of chemical proportion teach us that carbonic acid is composed of *one atom* of carbon vapor, (by weight 6,) and *two atoms* of oxygen, (by weight 16,) the latter, in volume, **being** double that of the former, as in the annexed figure :

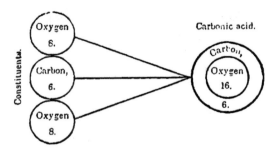

Thus, as far as the carbon is concerned, we obtain the infor mation we sought, namely, its saturating equivalent of oxygen and which we find to be just *double its own volume;* or, by weight, as 16 is to 6. But, without the aid of chemistry, we should here have remained satisfied; combustion would appear to have been complete; there would be no smoke, and no visi ble indication of an imperfect or unfinished process. Yet, chem istry tells us, we have only disposed of the *one* constituent of the gas, namely, the *carbon*, and that the *hydrogen, the second constituent*, remains yet to be accounted for, and converted to heating purposes. *

It is true, the carbon was, *in weight*, equal to six parts out of eight (the original weight of the gas.) *In bulk*, however, it was but one *fifth;* and when it is recollected, that, although the *illuminating* properties of the carbon are superior to those of the hydrogen, yet that the *heating* properties of the hydrogen are far superior to those of the carbon, we can appreciate the loss sus tained should these four fifths of the gas remain unconsumed.

To this may be added, the probable injury done to the heat ing powers of the flame by the conversion of any part of this otherwise valuable *hydrogen* into one of the most destructive compounds which can be met with in the furnace or flues,

---

* I have here stated the case of the oxygen uniting with the *carbon*, before the *hydrogen*. Chemists are undecided on this point; and, indeed, the evidence at present is quite contradictory.

It is to be observed, however, that the argument, drawn from the combustion of the carbon before the hydrogen, or *vice versa*, is the same, as regards the point now under consideration. Whichever half **passes off** uncombined, is lost.

namely, *ammonia*, composed of unconsumed hydrogen and a portion of the nitrogen liberated from the air. Thus we have a double motive for providing against the escape, *unconsumed*, of the hydrogen of the gas.

What, then, is to be done? Let us complete this *second process* as we did the first: let us supply this hydrogen, this remaining 80 per cent. in volume of the gas, with its own proper equivalent of oxygen, as we did in the case of the carbon.

But what is this second equivalent? By the same laws of definite proportions, we learn that the saturating equivalent of an atom, or any other given quantity of hydrogen, is, not *double* the volume, as in the case of the carbon, but *one half* its volume only — the product being aqueous vapor, that is, *steam ;* the relative weights of the combining volumes being 1 of hydrogen to 8 of oxygen; and the bulk, when combined, being two thirds of the bulk of both taken together, as shown in the annexed figure 8. *

We thus find, that to saturate the *one* volume of carbon vapor, two volumes of oxygen are required; whereas, to saturate the *two* volumes of hydrogen, *one* volume only of oxygen is required: thus,

FIRST CONSTITUENT.

Carbon.	Oxygen.	
Vol. Atom. Weight.	*Vol. Atom. Weight.*	*Vol. Atom. Weight.*
$\frac{1}{2}$ . .1 . . . .6 unite with	1 . .2 . . .16 forming carbonic acid.	1 . .1 22

SECOND CONSTITUENT.

Hydrogen.	Oxygen.	
Vol. Atom. Weight.	*Vol. Atom. Weight.*	*Vol. Atom. Weight.*
2 . .2 . . . .2 unite with	1 . .2 . . . .16 forming steam.	2 . .2 18

Here we see, that, in the case of this *first* constituent, as above, the *half* volume of carbon and *one* volume of oxygen

---

* Professor Brande puts this so clearly that I here give his own words : — "The simple ratio which the *weights* of the combining elements bear to each other involves an equally simple law in respect to combining *volumes*, where substances either exist, or may be supposed to exist, in the state of gas or vapor.

"Thus, water may be considered as a compound of 1 atom of hydrogen and 1 atom of oxygen, the relative weights of which are to each become condensed into one volume of carbonic acid (as shown in the last figure); and that, in the *second* constituent, the *two* volumes (meaning double bulk) of hydrogen, and *one* volume of oxygen, become condensed into two volumes of steam, (as shown in the annexed figure.)

other as 1 to 8. Hence, the *equivalent* of the atom of water will be, 1 hydrogen + 8 oxygen = 9. But oxygen and hydrogen exist in the gaseous state, and the weight of *equal volumes* of those gases (or, in other words, their relative densities, or specific gravities) are to each other as 1 to 16; hence, 1 volume of hydrogen is combined with ½ a volume of oxygen to form 1 volume of the *vapor of water*, or *steam*: for the specific gravity of steam, compared with hydrogen, is as 1 to 9. The annexed diagram, therefore, will represent the combining *weights* and *volumes* of the elements of water and of its vapor.''

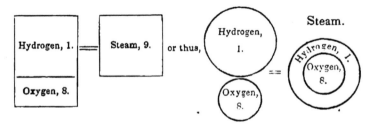

The following is also much to the point : — " La composition de l'eau est un des élémens les plus necessaires aux calculs des chemistes, les derniers experiences de MM. Berzelius et Dulong out fourni pour sa composition des nombres qui sont adoptés par tous les chemistes. Elle est formée d'apres eux de

Oxygène . . . . . 88.90 . . . . . . 1 volume, oxygène.
Hydrogène . . . . 11.10 . . . . . 2 volumes, hydrogène.
———— ——
100.00   1 volume eau.

Parmi les nombreuses decouvertes que la science doit a M. Gay Lussac, on remarquera toujours la belle observation sur la composition de l'eau, qui le conduisit a trouver les vrais rapports des gaz et des vapeurs dans leurs combinaison. Des experiences tres exactes, qu'il avoit faites conjointement avec M. de Humboldt, lui prouverent *que l'eau formée d'un volume d'oxygène et de deux volumes de hydrogène*, resultat plainement confirmé depuis par tous les phenomènes ou l'eau joue un role actif, et qui s'accorde avec la composition trouvé par MM. Berzelius et Dulong.''
— *Dumas*, vol. i., p. 33.

No facts in chemistry, therefore, can be more decidedly proved, than that one atom of hydrogen and one atom of oxygen (*the former being double the bulk of the latter*) unite in the formation of water; and, further, that one atom of carbon vapor and two atoms of oxygen (*the latter being double the bulk of the former*) unite in the formation of carbonic acid gas.

Thus, the ultimate fact of which we were in search is, that the one condensed volume of the gas, as generated from the coal, requires two volumes, or *double its bulk of oxygen*, that being the quantity required for the saturation of its constituents *when separated*.

Now, this is the entire alphabet of the combustion of the car-buretted hydrogen gas.

Having thus ascertained the quantity of *oxygen* required for the saturation and combustion of the two constituents of coal gas, the only remaining point to be decided is, *the quantity of air that will be required to supply this quantity of oxygen.*

This is easily ascertained, seeing that we know precisely the proportion which oxygen bears, in volume, to that of the air. For, as the oxygen is but *one-fifth* of the bulk of the air, *five* volumes of the latter will necessarily be required to produce *one* of the former; and, as we want *two* volumes of oxygen for each volume of the coal gas, it follows, that *to obtain those two volumes, we must provide ten volumes of air.*

Thus, then, by strict chemical proof, we have obtained these facts : — First, that each volume of coal gas requires two volumes of oxygen; secondly, that to obtain these two volumes of oxygen we must employ eight atoms of air; thirdly, that these eight atoms of air are equal to ten volumes of the coal gas; each volume of the latter, in fact, requiring ten volumes, or *ten times its bulk* of air : thus,

Ten *volumes* of air are the same as eight *atoms ;*
Eight atoms of air produce four atoms of oxygen ;
Four atoms of oxygen are equal to two volumes of the same ; and
Two volumes of oxygen saturate one volume of the coal gas :
Therefore, *ten* volumes of air are required for each *one* volume of this gas.

We now see why *ten* volumes of air are required for each

volume of gas, and why *neither more nor less* will satisfy the conditions of its combustion. For, if *more*, the excess, independently of the mischievous chemical unions that might enter into it in the furnace, would be the means of carrying away as much heat as it would take up by its expanding faculty. And if *less*, a corresponding quantity of either hydrogen or carbon would be deficient of its *supporter*, and necessarily pass off uncombined and unconsumed.

The only observation here necessary to make on the difference between these two gases is, that as this latter gas contains *two* atoms of carbon instead of *one*, it follows that a proportionate additional quantity of oxygen will be required for this additional atom of carbon. Hence, if carburetted hydrogen requires *two* volumes of oxygen for combustion, the bi-carburetted hydrogen will require *three* volumes. And so of air: if ten volumes of air are required for the one gas, fifteen volumes are consequently required for the other gas.

4. We have seen that, in the formation of the carburetted hydrogen, a considerable portion of the carbonaceous constituent of fuel is separated, and carried away by the hydrogen in the *gaseous form*, forming the carburetted hydrogen; the remainder of such carbonaceous matter is what we have now to deal with; the difference as regards combustion between these two portions of carbon being so important as to demand especial notice.

In observing this curious arrangement by which the saturation of the combustible atoms is effected, we perceive that *three* atoms of the combustible are apportioned to *four* of the supporter. This, we see, is the result of *one* atom of carbon requiring *two* of the supporters, while the two of hydrogen are satisfied with one each.

Now, in this arrangement no excess or deficiency appears among the heat-producing ingredients. Could we have dispensed with or avoided the presence of such an excess of nitrogen, (which is neither a combustible nor supporter of combustion,) the several unions would have been less embarrassed, — their combustion more rapid and complete, — and the intensity of their action much increased. That, however, was impossible.

The presence of so large a quantity of nitrogen being the una·
voidable condition of obtaining the oxygen through the instru·
mentality of atmospheric air,

It is to be observed that the process of combustion here
described is the most perfect that could be produced, either in a
furnace or lamp. Any deviation, therefore, by means of excess
or deficiency, or from any interruption or interference, such as
the interposition of another gas, must be more or less destructive
to the desired effect, viz., *the generation of the greatest quantity
of available heat.*

5. When we speak of mixing a given quantity of oxygen
with a given volume of smoke, (or coal gas,) we do so because
we know that such quantity of the former is required to saturate
the latter, and by such saturation every atom of *both* gases
enters into union, without excess or deficiency of either, producing entire and complete combustion.

So, when we speak of mixing a given volume of atmospheric
air with a given volume of smoke, we do so for the same purpose, knowing that the precise quantity of air will provide the
required quantity of oxygen.

Thus, if we know that *two* cubic feet of oxygen are the exact
saturating equivalent, or combining volume, for effecting the entire combustion of one cubic foot of coal gas, we know that *ten*
cubic feet of atmospheric air will effect the same purpose,
because ten cubic feet of air contain the required two cubic feet
of oxygen.

We require *ten* cubic feet of air to supply *two* cubic feet of
oxygen, which, if the air be pure, effects the combustion of *one*
cubic foot of coal gas, emanating from coals in the process of
combustion in a furnace; but if this quantity of air does not
contain this 20 per cent., or one-fifth, of oxygen, it is clear we
cannot obtain it. The air, in this case, may be said to be vitiated, or impure. It is therefore desirable that the air admitted
into a furnace should be direct from the atmosphere; otherwise,
the oxygen contained may be deficient, although the volume
of air admitted be sufficiently large.

Let us now inquire how far the ordinary mode of constructing and managing our furnaces enables us to satisfy this condition.

In ordinary furnaces, the supply of air is obtained by means of the *ash-pit ;* and the larger the ash-pit, the greater the quantity of air admitted. The ash-pit is made larger, under the mistaken notion that the more air we give, the better will be the draught, the more complete the combustion, and the greater the quantity of heat produced.

There can scarcely be a more absurd practice than is involved in this one-sided view of the principles of combustion, even supposing that the introduction of air is tantamount to the introduction of oxygen. It is manifest, however, that there are two different processes going on in the furnace, and two different combustibles, requiring their respective volumes of oxygen to consume them, namely, the gas or smoke generated in the body or cavity of the furnace, and passing off by the flues, and also, the solid carbon resting on the bars, both of which require separate volumes of oxygen to effect their combustion.

All that seems to be concluded in practice is, that air is essential to combustion ; and that if air be admitted to the fuel, through between the bars, it will work out the process of combustion satisfactorily in its own way. And hence the many errors and absurdities of the present system of practice.

There can be no greater mistake than letting a large quantity of air act directly on the burning fuel, which acts like a blast upon the red-hot mass, driving off the gases more rapidly, but also driving off the contained heat, and consuming the fuel with unnecessary rapidity.

It seems to be taken for granted, that if air, *by any means*, be introduced to the fuel in the furnace, it will, as a matter of course, mix with the gas, or other combustible, in a proper manner, and assume the state suitable for combustion, whatever be the nature or state of such fuel, and without regard to time or other circumstances. Now, it might as well be supposed, that by bringing large masses of nitre, sulphur, and charcoal together, we could form gunpowder. We know that it is by the proper mixture and incorporation of the different elementary atoms that simultaneous action is imparted to the whole ; and

13*

so, also, by bringing different kinds of gases into a state of preparation for simultaneous action.

The complete combustion of a body depends upon the chemical union of its atoms, or elementary divisions, with their respective equivalents of the *supporter*, oxygen; and which necessarily implies the *bringing together*, and the mixing of such atoms, previous to the mixture being fired for combustion.

It is not our purpose to enter upon the theory of atomic mixtures, or the time required to effect their combination, — which will be found in the numerous chemical works of the present day. We will now proceed to consider the means by which air may be introduced to the furnace, to effect the combustion of the gases therein generated.

In looking for a remedy for the evils arising out of the hurried state of things which the interior of a furnace naturally presents, and observing the means by which the gas is effectually consumed in the Argand lamp, it seemed manifest, if the gas in the furnace could be presented by means of jets to an adequate quantity of air, as it is in the lamp, the result would be the same, — namely, a quicker and more intimate mixture and diffusion, and consequently a more extensive and perfect combustion. The difficulty of effecting a similar distribution of the gas in the furnace, by means of jets, however, seems insurmountable. One alternative alone remains: since the gas cannot be introduced by jets into the body of the air, the air might be introduced by jets into the body of the gas; and this will be an effectual remedy.

Fig. 33 is a section of Williams' furnace for the prevention of smoke. In this furnace, the fuel, as will be seen from the cut, is thrown immediately upon the grate bars, and through them the air finds admission to it for the purpose of consumption. The gases pass over the bridge C; here they meet a current of air entering just beyond the bridge, which has been admitted by the air-tube *b*, below the ash-pit *f*, into the air-chamber *d*, and from thence escaping through a great number of small apertures in the diffusion plate above.

The force with which the air enters through this series of jets or blow-pipes enables it to penetrate into the gases, and

Fig. 33.

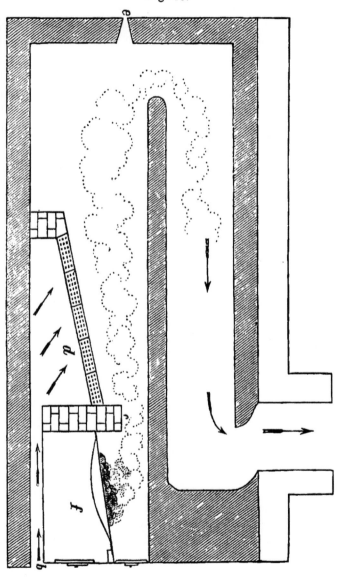

obtain the largest possible extent of contact-surfaces for the air
and gases; which is important, since the short time allowed for
the diffusion would otherwise be insufficient, in consequence of
the rapid passage of the smoke and gases over the diffusion
plates; *e* is the spy-hole for ascertaining the state of the smoke.

Fig. 34 is an apparatus invented by Mr. Jeffreys, of Bristol, as
long ago as 1824, for precipitating the lamp-black, metallic
vapors, and other sublimated matters from smoke, by washing
the latter by means of a stream of water. Where the necessary
supply can be secured, this plan is both effectual and economical, and well adapted for situations where the presence of smoke,
as well as the impurities produced by it, is an annoyance.

In the vertical section, B B is the smoke flue. The smoke
passing in the direction of the arrows at A, the flue turns down-
ward; and at the top of this vertical portion is a cistern E, the
perforated bottom of which lets down a constant stream of
water, after it is set to work. The shower, in its descent, carries
all the smoke and the sublimated matter which has passed from
the fire, which runs off at the bottom, F. The flue may then
turn upwards, or enter a common chimney; but little or nothing
will pass up it, providing the water be kept constantly running.
This apparatus is easily constructed, and is admirably suited for
hot-houses situated in the midst of pleasure-grounds, where
smoke is unsightly and disagreeable.

Whether these methods of consuming the gases generated in
the furnaces and flues of hot-houses may be considered worthy
of general adoption, we cannot tell. It is, nevertheless, presented to the consideration of the ingenious mechanic, not
doubting that were the subject fully taken up by energetic furnace builders, something good would be the result. That
immense quantities of fuel are wasted by imperfect combustion,
cannot be doubted, when we see the dense volumes of smoke
proceeding from chimneys where much heat is required.

Professor Brande says, " when air is admitted in front of the
furnace, or through or over the fuel, it obviously never can
effect those useful purposes, which are at once obtained by
admitting it in due proportion to the intensely heated inflammable vapors and gases, or, in other words, to the products of the

Fig. 34.

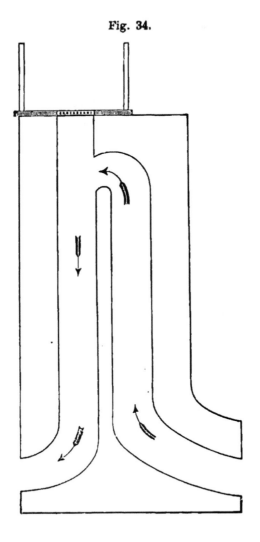

distillation of coal, at such temperatures that they may take fire in its contact." If a number of jets of air be admitted into a heated inflammable atmosphere, as the body of a furnace, its combustion will be attained in such a way as to produce a great increase of heat, and, as a necessary consequence, destroy the smoke.

In some of the large gardens of Europe, as well as in some manufactories, attempts have been made to consume the smoke or gases of the furnaces, by bringing them in contact with a body of glowing incandescent fuel, producing a result the reverse of what was expected, namely, the absorption of heat by their expansion and decomposition, instead of giving out heat by their combustion. It is strange that this erroneous notion should be persisted in, even at the present day, when any chemical work of good authority would satisfy any one wishing for such knowl-edge that *decomposition*, not *combustion*, is the effect of a high temperature being applied to hydro-carbon-gases; — that no possible degree of heat can consume *carbon ;* — that it is a well-known property of both the varieties of carburetted hydrogen, that they deposit charcoal, (carbon) virtually become *smoke*, when heated; — that the amount of carbon deposited is propor-tioned to the increase of temperature, and that its *combustion* is merely produced by, and is, in fact, its *union* with, oxygen, which these smoke-burners take no care to provide.*

---

* Numerous methods have been devised for *burning* smoke, and patents have been issued for supposed inventions of this kind, showing the want of chemical knowledge on this subject. One consists in hav-ing a double set of fire-bars, so that when the fuel is red-hot, it is thrown back on the innermost bars, and the smoke of the fresh coal in front passing over this incandescent fuel, is supposed to be consumed in its passage. Another proposes a sliding carriage for this purpose, working on castors inside the furnace. Others of a similar kind have been put forward, and all on the same principle; all manifesting the same neglect or ignorance of chemistry, — for chemistry teaches us that *heat* has nothing to do with the *combustion* of smoke beyond this, — that a certain temperature is essential to the development of chemical action between the combustible and the supporter, when they are brought together. But producing heat is not producing air; and decom position is not, in this respect, combustion.

The neglect of chemistry when treating of combustion, and the results of this neglect in these *smoke-burning* furnaces, cannot be too strongly exposed; neither can its study be too strongly enforced, seeing that it is practically within the reach of all. For chemistry is no longer the mysterious alchemy that it was a century ago; it is now a mere rigid inquiry into nature's processes and laws, by the aid of those proofs and illustrations which nature herself has supplied. It has taken its place among the exact sciences, and now recognizes no man's dictum or opinion, apart from experimental tests, and strict, substantial evidence.

Looking, then, to chemistry, we would add, in reference to these smoke-burning expedients, that, in seeking to obtain heat from gas, (or smoke,) the bringing it into connection with ignited carbonaceous matter, or to anything approaching the temperature of incandescence, is absolutely useless, if not injurious, until we are assured of having the means of contact with air fully provided for.

The mere enunciation of a plan "for *consuming smoke*," is *prima facie* evidence that the inventor has not studied and considered the subject in its chemical relations. Chemists can understand a plan for the *prevention* of smoke; but as to its *combustion*, it is so unscientific, not to say impossible, (if there be any truth in chemistry,) that such phraseology should be avoided. The popular phrase, "A furnace burning its own smoke," may be justifiable, as conveying an intelligible meaning; but, in a work having any pretensions to science, or from any one pretending to teach those who are unable to distinguish for themselves, and who may easily be led into error, is wholly objectionable.

6. *Construction of Furnaces.* — From what has been already said, in the preceding part of this section, it will be seen that the construction of furnaces is a matter of great importance in the economy of heat. To investigate the various varieties of furnaces which have been recommended, would occupy too much of our space at present, especially as we shall have to refer to them hereafter, when treating of the different methods of heating; besides, in small apparatuses, the intense heat required

for large boilers is unnecessary. A very moderate heat, applied on the most economical principle, and the furnace so constructed as to make the fuel burn for a long time, without much attention, and without much escape of smoke, is the grand desideratum, and which is easily accomplished, with a moderate degree of care and skill in the erection.

Passing over, then, as unnecessary for our purpose at present, the many ingenious forms which have been given to furnaces, we will proceed to describe the most simple plan, which, in our experience, is the most effectual in the combustion of the fuel, as well as the least expensive in the construction.

It should be an object of consideration, in building the furnace, to confine the generated heat within the cavity of the furnace as much as possible, so that the gases generated by the combustion of fuel may be prevented from passing too rapidly along the flue; this is more especially requisite with boiler furnaces. The throat of the furnace should be contracted as much as possible. In furnaces where the only entrance for air is by the bars, provision should be made for the entrance of enough — but no more than enough — for the combustion of the fuel, and the entrance should, in all cases, be regulated by a damper, on the ash-pit door. It should be considered, that the rarity of the heated gases causes them to force their passage through the throat of the furnace, just in the proportion of its size.

We have already shown that any air entering through the door of the furnace reduces the intensity of the heat, although it is supposed by some that the passage of air over the burning fuel promotes the more perfect combustion of the gaseous products of the coal. But even if this be correct, the heat will be reduced, and less heat will be generated in a given time, than if the whole gaseous products escaped by the chimney.

The kind of fuel to be burnt must, in all cases, determine the width of the bars; and as a certain open area is necessary for the admission of air to effect combustion, it is desirable that this area should be known.

Supposing the ordinary kind of furnace bars to afford about thirty inches of opening for air for every square foot of surface, —

then supposing you wish to erect a hot water apparatus — the relative proportions between the area of the bars and the length of pipe would be as follows : —

Area of Bars.					4 *inch pipe.*	3 *inch pipe.*	2 *inch pipe.*
75 square inches will supply					150 feet, or	200 feet, or	300 feet.
100	"	"	"	"	200 "	266 "	400 "
150	"	"	"	"	3C0 "	400 "	600 "
200	"	"	"	"	400 "	533 "	800 "
250	"	"	"	"	500 "	666 "	1000 "
300	"	"	"	"	600 "	800 "	1200 "
400	"	"	"	"	800 "	1066 ·"	1600 "
500	"	"	"	"	1000 "	1333 "	2000 "

Thus, suppose there are six hundred feet of pipe, four inches in diameter, in an apparatus, — then the area of the bars should be three hundred square inches, so that thirteen inches in breadth, and twenty-three inches in length, will give the required quantity of surface. When it is required to obtain the greatest heat in the shortest time, the area of the bars may be a little increased.

In order to make the fire burn for a long time without attention, the furnace should extend beyond the bars, both in length and breadth ; and the coals, which are placed on this blank part of the furnace, in consequence of receiving no air from below, will burn slowly, and will only enter into complete combustion when the rest of the coal, on the bars, has been consumed.

It may be observed, that as the maximum effect of the furnace is seldom required, the register on the ash-pit door, and the damper in the flue, must be used to regulate the draught, and **thus** limit the consumption of fuel.

14

PRINCIPLES OF HEATING HOT-HOUSES.

1. *Effects of artificial heat.* — The effects that are produced upon the functions of vegetables, by atmospheric air that has passed over intensely heated surfaces, are perceptible to the most casual observer. The changes, therefore, that are produced upon atmospheric air by subjecting it to a high temperature, are of the utmost importance to the horticulturist, and consequently demand our particular attention.

When common air passes over highly heated surfaces, the small particles of animal and vegetable matter, (organic matter,) which are always held in suspension by it, are decomposed by the heat, and resolved into various elementary gases. This is one of the causes of the unpleasant smell which results from this method of heating, as in common stoves, Polmaise furnaces, &c. But, in addition to this, the aqueous vapors of the atmosphere are almost entirely decomposed, the oxygen entering into combination with the iron, and the hydrogen mixing with the air. The changes which have thus taken place, render the atmosphere extremely deleterious to both animal and vegetable life.

The mixture of the hydrogen thus disengaged is even more injurious to the plants than the alteration which has taken place in its hygrometric state, as this will be partly supplied by the moisture contained in their tissue, until it be restored to the atmosphere by evaporation, which is easily effected.

The particles of animal and vegetable matter — as we have said — are decomposed by the heat; and they then produce extraneous gases, consisting of sulphuretted, phosphuretted, and carburetted hydrogen, with various compounds of nitrogen and carbon, which, in the state in which they exist, are highly inimical to vegetable life. *

The quantity of hydrogen which is eliminated by the decomposition of water contained in the air is one thousand three hundred and twenty-five cubic inches for every cubic inch of water that is decomposed; and if the dew point of the air be 45° at an average, this quantity will be given out from every seventy-two cubic feet of air which passes over the heated surface. It is, therefore, not difficult to account for the effects produced on vegetation by hot-air stoves, in consequence of the air, when thus artifically dried, abstracting too much moisture from their leaves. It is also clear that the injury must increase in proportion to the length of time the apparatus continues in use, by the plants being surrounded by, and compelled to inhale, these extraneous gases, which are evolved from the decomposition of the constituents of the atmosphere.

The extreme dryness of the air, after it has been deprived of

---

* I am unable to ascertain the exact nature and extent of the change which atmospheric air undergoes by being passed over intensely heated metallic bodies; but whatever be the chemical alteration which occurs, a physical change undoubtedly takes place, by which its electrical condition is altered.

From some experiments recorded in the *Philosophical Transactions* of the Royal Society, made with a view of ascertaining the effect produced on the animal economy by breathing air which has passed through heated media, it appears that the air which has been heated by metallic surfaces of a high temperature must needs be exceedingly unwholesome. A curious circumstance is related, in reference to these experiments, which is illustrative of this fact.

"A quantity of air, which had been made to pass through red-hot iron and brass tubes, was collected in a glass receiver, and allowed to cool. A large cat was then plunged into this air, and immediately she fell into convulsions, which, in a minute, appeared to have left her without any signs of life; she was, however, quickly taken out and placed in the fresh air, when, after some time, she began to move her eyes, and, after giving two or three hideous squalls, appeared slowly to recover. But on any person approaching her, she made the most violent efforts her exhausted strength would allow to fly at them; insomuch, that, in a short time, no one could approach her. In about half an hour she recovered, and became as tame as before."

its hygrometric vapor by passing over a hot-air stove, such as polmaise, is productive of the worst consequences to growing plants. To remedy this evil, a trough of water is laid over the heating surface, which in some degree mitigates this evil. The evil, however, cannot be entirely got rid of by this means; for even if the proper quantity of moisture can be again restored to the air, the effects which result from the use of extraneous gases are in no way removed. When the surface of radiation is an iron plate, these injurious effects are much greater.

The heating by means of brick flues is, in some respects, similar to the effects produced by hot-air stoves, but only when the flues are heated to a high temperature, which is unnecessary. In the latter case, an unwholesome smell is also produced, by the decomposition of the organic matter in the atmosphere, and in some cases, probably, by a small portion of sublimed sulphur from the bricks, as well as by the escape of various gases through the joints or accidental fissures of the flues. These contingent causes may, however, be in a great measure avoided. The hygrometric vapors of the atmosphere are not decomposed by this system of heating, as by a hot-air stove, because when the flues are warmed to a common temperature, the heat is perfectly pure, and the materials of which the flues are built having but little affinity for oxygen, they are consequently more healthy than hot-air stoves.

Air passing over a highly heated surface of *iron* is, therefore, more injurious than when passed over any other body, as stone, or brick, as the power of iron to decompose water increases with the temperature to which it is heated. The limit to which the temperature of any metallic surface ought to be raised, for warming horticultural buildings, (or indeed any other buildings,) is 212°, if a healthy, uncontaminated atmosphere be desired. The importance of this rule cannot be too strongly insisted on, for upon it entirely depends the healthiness of every system of artificial heat.

2. *Laws of Heat.* — Heated bodies give off their caloric by two distinct methods — *radiation* and *conduction*. These are governed by different laws; but the rate of cooling — or parting

with heat — by both modes, increases in proportion as the heated body is of greater temperature above the surrounding medium.

The cooling of a heated body, under ordinary circumstances, is evidently the combined effects of radiation and conduction; the conductive power of the air is, evidently, owing to the extreme mobility of its particles, for otherwise it is one of the worst conductors with which we are yet acquainted, so that when confined in such a manner as to prevent its freedom of motion, it becomes useful as a non-conductor.

The proportion which radiation and conduction bear to each other has, in general, been very erroneously estimated. Count Rumford considered the united effect, compared with radiation alone, was as five to three, and Franklin supposed it to be as five to two.

No such general law, however, can be deduced, for the relative proportions vary with the temperature, and with the peculiar substance, or surface, of the heated body; for, while the cooling effects of the air, by conduction, is the same on all substances, and in all states of the surface of those substances, radiation varies very materially, according to the nature of the surface.

The influence of the air, by its power of conduction, varies also with its elasticity. The greater its elastic force, the greater also is its power of cooling, according to the following law: — When the elasticity of the air varies in a geometrical progression whose ratio is 2, its cooling power also changes in a geometrical progression whose ratio is 1.366.

The same law holds with all gases, as well as with atmospheric air; but the ratio of the progression varies with each gas.

To show the relative velocities of cooling at different temperatures, the following table, constructed from the experiments of Petit and Dulong, is given. The first column shows the excess of temperature of the heated body above the surrounding air; the second column shows the rate of cooling of a thermometer with a plain bulb, and the third column gives the rate of cooling when the bulb was covered with silver leaf. The fourth column shows the amount due to the cooling of the air *alone;* and by deducting this from the second and third columns respectively

14*

we shall find what is the amount of radiation under the two different *states of surface*, noticed at the top of the second and third columns. *

Excess of temperature of the thermometer above that of the air. Centigrade Scale.	Total velocity of cooling of the naked bulb.	Total velocity of cooling of the bulb covered with silver leaf.	Amount of cooling due to conduction of air alone.
260°	24·42	10·96	8·10
240°	21·12	9·82	7·41
220°	17·92	8·59	6·61
200°	15·30	7·57	5·92
180°	13·04	6·57	5·19
160°	10·70	5·59	4·50
140°	8·75	4·61	3·73
120°	6·82	3·80	3·11
100°	5·57	3·06	2·53
80°	4·15	2·32	1·93
60°	2·86	1·60	1·33
40°	1·74	·96	·80
20°	·77	·42	·34
10°	·37	·19	·14

Some very remarkable effects may be perceived by an inspection of the above table. It appears that the ratio of heat lost by contact of the air alone, is constant at all temperatures; that is, whatever is the ratio between 40° and 80°, for instance, is also the ratio between 80° and 160°, or between 100° and 200°. This law is expressed by this formula:

$$v = n.\, t^{\,1\cdot33};$$

where t represents the excess of temperature, and n a number which varies with the size of the heated body. In the case represented in the foregoing table, $n = 0.00857$.

Another remarkable law, is that the cooling effect of the air is the same, for the like excess of heat, on all bodies, without regard to the particular state or nature of their surface. This

* The temperatures of this table are expressed in degrees of the Centigrade thermometer, as the zero of this thermometer is the freezing point of water, and from that to the boiling point of the same fluid is 100°. In order to find the number of degrees on Fahrenheit's scale, which answers to any given temperature of the Centigrade, multiply the number of degrees of Centigrade by 9, and divide the product by 5; add 32 to the quotient thus obtained, and this sum will be the number of degrees of Fahrenheit required.

was ascertained by Petit and Dulong, in a series of experiments, not necessary here to detail, but which proved the accuracy of the deduction.

By comparing the second and third columns of the above table, it will be immediately perceived that the loss of heat by *radiation* varies greatly, with the nature of the radiating surface; though, whatever be the nature of the surface, the *loss of heat is the same in all cases*, though in a different ratio.

It should be observed, that, in this table, the second, third, and fourth columns show the number of degrees of heat which were lost per minute by the body which was subject to the experiment; and, therefore, these numbers represent the velocity of cooling.

The fact, already adverted to, that the ratio of cooling in those bodies that radiate least is more rapid at low temperatures, and less at high temperatures, than those bodies that radiate most, is, perhaps, one of the most remarkable of the laws of cooling. It was first deduced experimentally by Petit and Dulong, and it may be mathematically proved from their formula; but it is unnecessary here to enter into the investigation. It appears, however, that when the total cooling of two bodies is compared, the law is more rapid at low temperatures for the body which radiates least, and less rapid for the same body at high temperatures; though separately, for conduction and radiation, the law of cooling is, for the former, irrespective of the nature of the body, and for the latter, that all bodies preserve at every difference of temperature a constant ratio in their radiating power.

It is not our purpose to enter minutely into detail on the laws of heat, which will be found in modern works on chemistry, and which ought to form part of the studies of all young gardeners who wish to become acquainted with the principles of hot-house management. We will now proceed to consider the specific properties of air and water as agents in the heating of horticultural structures.

3. *Specific heat of air and water.* — Very erroneous notions are entertained by many persons as to the absolute quantity of

heat taken up by different substances. To ascertain, therefore, the effect a certain quantity of water will produce in warming the air of a hot-house, there appears to be no better method than that of computing from the specific heat of gases compared with water.

Every substance has its peculiar specific heat. Now, one cubic foot of water, by losing one degree of heat, will raise the temperature of 2990 cubic feet of air the extent of one degree; and, by the same rule, by losing 10° of its heat, it will raise the temperature of 2990 cubic feet of air 10 degrees; and so with similar quantities in similar proportions.

In order to know the time it will take to heat a certain quantity of air any required number of degrees, by means of hot water contained in metal pipes, we must calculate the effect from direct experiment; and, as the radiating and conducting powers of different substances differ considerably, it is necessary that the experiment be made with the same material as the pipes for which we wish to estimate the effect.

From data obtained by experiments on the cooling of iron pipes, it appears that the water contained in a pipe 4 inches in diameter loses ·851 of a degree of heat per minute, when the excess of its temperature is above 125 degrees above that of the surrounding air. There one foot in length of a pipe 4 inches diameter will heat 222 cubic feet of air one degree per minute, when the difference between the temperature of pipe and the air is 125 degrees.

To calculate from this data, however, the length of a pipe, of any given size, that will be necessary to warm a house, and to maintain it at any given temperature under a certain external temperature, it will be necessary to estimate the heat lost by the conducting and radiating power of the glass, and of any metallic substance used in the structure.

Heating horticultural structures is a very different matter from heating solid opaque buildings; and here many erectors of heating apparatus fall into error. They suppose, because an apparatus of certain power heated a large building, — a church or a hall, — one of proportionate dimensions should warm a hot house of proportionate size, without taking into full considera tion the great difference of the external radiation, and the conduction of heat by the materials of the building.

The loss of heat by buildings covered with glass is very great. It appears, by experiment, that one square foot of glass will cool down 1·279 cubic feet of air as many degrees per minute as the internal temperature of the house exceeds the temperature of the external air; thus, if the difference between the external temperature and the temperature of the house be 30 degrees, then 1·279 cubic feet of air will be cooled 30 degrees by each square foot of glass; or, more correctly, as much heat as is equal to this will be given off by each square foot of glass, for, in reality, a very much larger quantity of air will be affected by the glass, but it will be cooled to a less extent. The real loss of heat, however, from the house will be what is here stated.

There are various causes likely to affect these calculations, such as, —

High winds, which are found to reduce the internal temperature more than actual cold, or even frost;

Condensation of moisture on the glass, which prevents the escape of heated air; and, when a certain temperature is maintained within, prevents radiation from the glass to a great degree;

The extent of wood in the roof of the house, which also prevents radiation and conduction, as in the case of metallic roofs.

These circumstances will be found to affect, in a greater or less degree, the air of the house, though, under general circumstances, these calculations will be nearly correct.

In estimating the quantity of glass surface contained in a building, the extent of wood surface must be carefully excluded. This is particularly necessary in all horticultural buildings, where the maximum of heating power is dependent upon the estimate taken. The readiest way of calculating, and sufficiently accurate for ordinary purposes, is to take the square surfaces of the sashes, and then deduct one eighth of the amount for wood work. In the generality of horticultural buildings, the wood work fully amounts to this quantity. When the frames and sashes are made of metal, the radiation of heat will be quite as much from the frame as from the glass; therefore no **deduction** is required in such cases.

From the preceding calculations the following corollary **may** be drawn: —

The quantity of air to be warmed *per minute* in habitable rooms and public buildings, must be 3½ cubic feet for each person the room contains, and 1½ cubic feet for each square foot of glass.

For conservatories, forcing-houses, and all buildings of this description, the quantity of air warmed *per minute* must be 1¼ cubic feet for each square foot of glass the structure contains.

When the quantity of air required to be heated has thus been ascertained, the length of pipe to heat it by hot water may be found by the following table:

Table of the quantity of pipe 4 inches diameter which will heat 1000 cubic feet of air per minute, any required number of degrees. The temperature of the pipe being 200° Fahrenheit: —

Temperature of external air.	Temperature at which the house is required to be kept.									
Fahrenheit's scale.	45°	50°	55°	60°	65°	70°	75°	80°	85°	90°
10°	126	150	174	200	229	259	292	328	367	409
12°	119	142	166	192	220	251	283	318	357	399
14°	112	135	159	184	212	242	274	309	347	388
16°	105	127	151	176	204	233	265	300	337	378
18°	98	120	143	168	195	225	256	290	328	368
20°	91	112	135	160	187	216	247	281	318	358
22°	83	105	128	152	179	207	238	271	308	317
24°	76	97	120	144	170	199	229	262	298	337
26°	69	90	112	136	162	190	220	253	288	327
28°	61	82	101	128	154	181	211	243	279	317
30°	54	75	97	120	145	173	202	234	269	307
Freezing point 32°	47	67	89	112	137	164	193	225	259	296
34°	40	60	81	104	129	155	184	215	249	286
36°	32	52	73	96	120	149	175	206	239	276
38°	25	45	66	88	112	138	166	196	229	266
40°	18	37	58	80	104	129	157	187	220	255
42°	10	30	50	72	97	121	148	178	210	245
44°	3	22	42	64	85	112	139	168	200	235
46°		15	34	56	79	103	130	159	190	225
48°		7	27	48	70	95	121	150	181	215
50°			19	40	62	86	112	140	171	204
52°			11	32	54	77	103	131	161	193

To ascertain, by the above table, the quantity of pipe required **to heat** 1000 cubic feet of air per minute, find, in the first column the temperature which corresponds to that of the external air, which may be the medium (or average) of your locality. Then, in the other column, find the temperature required in the house; then, in this latter column, and on the line which corresponds with the external temperature, the required number of feet of pipe will be found.

Supposing, now, that a forcing-house is to be kept at 75 degrees, and the average of the external thermometer in the coldest weather, taken at 10° (Fah.); then, by the foregoing table, we find, under the column 75°, and on the line 10°, for external temperature, the quantity 292, which is the number of feet of pipe required to heat 1000 cubic feet of air per minute, the proposed number of degrees. Of course, the volume of air in the house must be previously ascertained. Any other difference of temperature may be found in the same way.

It will thus be perceived, that the amount of heat required for warming a glazed structure is much greater than that required for warming an opaque building of the same size; in consequence of the radiation of heat from its surface; and the difference is much greater than the allowance made by erectors of heating apparatuses, under general circumstances.

To ascertain the effect of glass windows in cooling the atmosphere of a house, the following experiments were made, with a vessel as nearly as possible the same thickness as the glass ordinarily used for glazing. The temperature of the house, in these experiments, was 65°; the thickness of the glass was .0825 of an inch; the surface of the vessel measured 34·296 square inches, and it contained 9·794 cubic inches of water. The time in which this vessel cooled, when filled with hot water, is shown as follows : —

Thermometer cooled.		Observed time of cooling.	Calculated time of cooling.	Average rate of the observed time of cooling.
from	to			
150°	140°	6' 40"	6' 54"	1·176° per minute,
150	130	14 50	14 43	at an excess of 65°
150	120	23 30	23 40	above the tempera-
150	110	34 0	34 0	ture of the air.

From the average rate of cooling here given, the effect of glass in cooling the atmosphere of a room may easily be calcu lated, as the specific heat of equal volumes of air and water is as 1 to 2990. The above average will show that each square foot of glass will cool 1·279 cubic feet of air one degree per minute, when the temperature of the glass is one degree above that of the external air.

But by this we can only find the effect of glass in a still atmosphere, and, therefore, to find the effect of glass in cooling the volume of a hot-house, especially when exposed to the action of winds, further experiments are necessary, of which we shall treat in a subsequent part of this work, in connection with "protection of hot-house roofs during the night."

SECTION III.

HEATING BY HOT WATER, HOT AIR, AND STEAM.

1. THE practice of employing hot water, circulating through metallic tubes, or wooden troughs, for diffusing artificial heat in horticultural structures, though of recent origin, has now become so general, that its merits are fully acknowledged as the best method that has yet been invented, to effect the purpose with efficiency and economy. Until the last few years, — although its powers and properties were fully known, — it had been chiefly confined to a few cases of experiment, rather than to any general or useful purpose.

The present day, however, has fully revealed its merits, and shown the great, the unlimited, extent of its practical application and general utility. When we see such an immense structure as the great Palm house, lately erected at Kew Gardens, in London, heated with hot water in preference to all other modes; when we see the lately applauded mode of heating by steam abandoned; when we see the powerful, but unsuccessful, attempt to establish a new system of heating by hot air, called Polmaise, by some of the first horticulturists of England; when we see this system, notwithstanding its powerful supporters, driven into obscurity, and all but annihilated, by the well-tried superiority of hot water, which maintains its proud preëminence over all other methods of heating, and has its superiority acknowledged, even by its enemies.

One of the greatest advantages which this mode of heating possesses over all others, is, that a greater permanency of temperature can be obtained by it, than by any other method. The difference between an apparatus heated by hot water, and one heated by steam, is not less remarkable, in this particular, than in its superior economy of fuel.

15

2. *Comparison of heat in water and steam.* — The heating of
horticultural buildings by steam had its day and its admirers,
though both are now numbered among the things that were.
Even if the original outlay were equal, the additional outlay for
fuel, the risk of explosion from neglect, and the want of permanency in the apparatus to maintain the heat for any length of
time, are insuperable' objections to its adoption. Among many
instances that could be given of this method of warming large
houses, we might mention the large Palm house, in the Royal
Botanic Garden of Edinburgh, which was erected when heating
by steam was in the height of its fame. This house is about
fifty feet high and seventy-five feet wide, in the form of an
octagon ; the pipes are laid around the side of the wall. There
is a contrivance, however, resorted to here, in connection with
the system, to which its success in heating the house may be
somewhat, if not entirely, attributed. The steam is thrown into
large iron boxes, loosely filled with stones and pieces of brick,
for the retention and absorption of the heat. These iron boxes
are placed underneath the shelf that surrounds the house, and
close by the side of the wall, and at regular distances from
each other. By this contrivance, the temperature of the house
is kept up for a considerable time longer than would be by the
circulation of the steam alone. Indeed, we believe it was found
perfectly impracticable to maintain the proper temperature, during cold nights, until this expedient was adopted, viz., of filling
the boxes with absorbing materials.

We have known conservatories, in which steam apparatuses
had been erected, taken down, and their place supplied with
others of hot water, merely in consideration of the consumption
of fuel and extra attention required by a steam apparatus, keeping the danger of explosion out of the question.

It seldom happens that the pipes of a hot-water apparatus can
be raised to so high a temperature as 212° ; in fact, it is not
desirable to do so, because it is unnecessary to generate steam,
which would only escape by the air vent, without affording any
available heat. Steam pipes, on the contrary, must always be
above the temperature of 212°, otherwise steam will not be generated : and here the grand point to be attended to in artificial

heating is nullified, namely, the diffusion of heat at a low temperature. A given length of steam pipe, however, will afford more heat than one heated by hot water, by the aggregate calculation of its specific heat. But, if we consider the relative permanency of temperature, we shall find a very remarkable difference in favor of pipes heated by hot water; and the calculations here given are fully confirmed by experience and observation.

The weight of steam, at the temperature of 212°, compared with the weight of water at 212°, is about as 1 to 1694, so that a tube that is filled with water at 212° contains 1694 times as much *matter* as one of equal size filled with steam. If the source of heat be withdrawn from the steam pipes, the temperature will soon fall below 212°, and the steam immediately in contact with the pipes will condense; but, in condensing, the steam parts with its *latent heat*, and this heat, in passing from the latent to the sensible state, will again raise the temperature of the pipes; but, by the withdrawal of the heat from the boiler, the action of the cold air on the pipes quickly condenses the whole of the steam contained in them, which, when condensed, possesses just as much heating power as the same bulk of water at a similar temperature. This water now occupies only $\frac{1}{1694}$ part of the space which the steam originally did in the pipes.

The specific heat of uncondensed steam, compared with water, is, for equal weights, as ·8470 to 1; but the latent heat of steam being estimated at 1000 degrees, we shall find the relative heat obtainable from *equal weights* of condensed steam and of water, reducing both from the temperature of 212° to 60°, to be as 7·425 to 1; but for *equal bulks* it would be as 1 to 228; that is, bulk for bulk, water will give out 228 times as much heat as steam, reducing both to the temperature of 60°. A given bulk of steam, therefore, will lose as much of its heat in one minute, as the same bulk of water will lose in three hours and three quarters.

It must be considered, however, that when the water and steam are both circulated in iron pipes, the rate of cooling will be somewhat different from this ratio, in consequence of the much larger quantity of heat contained in the metal, than in the steam with which the pipe is filled.

The specific heat of cast iron being nearly the same as water, the water being 1000 and the iron 1100, if we take two similar pipes, four inches in diameter and one fourth of an inch thick, the one filled with water and the other with steam, each at the temperature of 212°, the one which is filled with water contains 4·68 times as much heat as the one which is filled with steam. Therefore, if the pipe with the steam cools down to the temperature of 60° in one hour, the one filled with water would require four hours and a half, under the same circumstances, before it reached the like temperature.

But this is merely reckoning the effect of the pipe and the fluid contained in it. In a steam apparatus, this is all that is effective in giving out heat; but in a hot-water apparatus there is likewise the heat from the water contained in the boiler, and even of the brick-work around the boiler, all which tends to increase the heat of the pipes, long after the fire is extinguished. In the one, the heat will continue to circulate through the pipes as long as any heat remains about the fire-place, because the circulation will continue in the pipes until the whole apparatus is cooled down. But, in the case of steam pipes, as soon as the water in the boiler falls below the boiling point, (212°,) circulation ceases, and the pipes then begin to cool, the remaining heat in the boiler and furnace goes for nought.

From these causes the difference in permanency of hot water and steam will be clearly apparent, and the fact of a house heated with hot water keeping up its temperature at least six times as long as one heated with steam, will be fully understood by those interested in the matter. These considerations are of the utmost importance to those erecting horticultural buildings, or, indeed, any other kind of buildings requiring artificial heat. This admirable property, which water possesses, of retaining its heat, of carrying it to any distance, and, without difficulty, giving it out gradually, or retaining it for many hours, renders it of vast importance to gardeners, and prevents the necessity of that constant attention to the fire, which forms so serious an objection in all other methods of heating.

We find, by experience, that no system of heating horticultural buildings in all respects answers the purpose so well as a
hot-water apparatus, well constructed, and judiciously arranged,
in regard to the amount of work it has to do, so that it may not
be necessary to strain it, on exigencies, to its maximum point of
strength. In whatever point of view it may be regarded, it is,
undoubtedly, the best for all practical purposes ; and the best
possible evidence of its utility is derived from the fact, that no
case has ever come under our knowledge, wherein it has failed
to give complete satisfaction, when it has been properly constructed, rightly managed, and judiciously arranged, in regard
to supplying a sufficient amount of radiating surface for the
work it has to do.

3. *Comparison of hot air with hot water, as a mode of heating
horticultural structures.* — Various erroneous opinions and principles have been theoretically and practically promulgated, in
regard to hot-air heating ; and, carrying with them, in general,
some degree of plausibility, and in some cases emanating from
men of learning, have led many, who have not studied the matter attentively, into very great errors. However invidious,
therefore, may be the task of pointing out such errors, we consider it our duty, when treating on the subject at large, not only
to exhibit what we consider to be the true principles, but to
show where erroneous principles have been adopted. This must
serve as an apology for the freedom with which the advocates
of Polmaise, and other methods of hot-air heating, and the systems they approve, are descanted on in this section.

We have already observed that the cooling of a heated body,
under ordinary circumstances, is evidently the combined effect
of radiation and conduction. The conductive power of the air
is principally owing to the mobility of its particles, for, otherwise,
it is one of the *worst conductors we are acquainted with.*

Atmospheric air, in passing into a house over a highly heated
surface, must necessarily lose a large quantity of its contained
moisture ; [see 1. Effects of Artificial Heat, of the preceding
section ;] and, as its capacity for taking up moisture is increased
according to its temperature, it follows that a great demand

15*

must be made upon the moisture of the house, upon the plants, and upon everything else within its influence capable of giving off moisture. This is also the case with hot-water pipes. But here the advantage of the latter is plainly illustrated ; for while a hot-air stove abstracts the moisture, in excess, from that part of the house nearest to the aperture of ingress, hot-water pipes radiate the heat at a *low temperature* equally over the whole surface, and, as the temperature at which the heat is radiated is comparatively low, little or no moisture is abstracted. Some suppose that they get a fine moist heat from hot-water pipes. This, however sound and sensible it may appear, is, nevertheless, a practical fallacy, the fact of the case being this, — that, instead of the moisture of the house being taken up by the air, as in the case of Polmaise, and other stoves, the warm air of the pipes being so much lower in temperature than that of the stoves, it cannot take it up, and hence the moisture remains with the plants and the atmosphere in its original purity. In fact, there is no difference between the heat radiated from stone, brick, or iron, unless it be mixed with extraneous gases, by heating these bodies to a high temperature.

To supply the moisture required by the heated air, water may be placed in evaporating pans, in connexion with the current of ingress ; but, as we have already shown, though moisture may be supplied, the hydrogen of the rarefied air still remains uncombined, and, until the air be replaced by a fresh volume from the external atmosphere, its impurities still remain.

With regard to the motion and circulation of the atmosphere of a hot-house, the system of heating by hot air possesses, theoretically, some advantages over all others. Strictly speaking, however, this has scarcely a practical foundation. If hot air be admitted in currents, the atmosphere will be agitated, certainly, but the house will be very unequally heated, as the heated air will pass upward in currents, at the aperture of its entrance, without diffusing itself over the lower surface of the house. Air expands, when heated, $\frac{1}{480}$ of its bulk for each degree of Fahrenheit, and the velocity of its motion is equal to the additional height which a given weight of heated air must have, in **order** to balance the same weight of cold air ; and as all **rare** bodies tend to rise vertically, in a dense medium, it follows, that when heated air enters a house by an aperture at one part of it, a very large portion of the heated air thus entering, must rise immediately towards the roof; and in practice we find this to be exactly the case. For, let any person examine the roof of a hot-house in a frosty night, heated by a hot-air stove, and he will perceive the part immediately above the entrance of the air quite warm by the ascending heat, while all the rest of the roof may be covered with ice or snow.

But the atmosphere of a house heated artificially, by whatever means, is always in motion; with hot-water pipes it may be less perceptible, for the reasons already stated, but it is not the less real. The motion given to the atmosphere of a house depends upon the difference of temperature between the two bodies of air, externally and internally; therefore a motion must continue in the air of a house artificially warmed, so long as the house requires warming, — that is, as long as any difference exists between the internal and external atmospheres.

Some advocates of hot-air heating found their arguments upon the fact that air can be raised to a higher temperature, in a given time, by a given amount of caloric, than water. This is probably true, if we calculate according to the bulk, without regard to the density, of the respective bodies; but, *supposing it to be true*, then we know that, by the law already referred to, *its rapidity in warming will just be in exact proportion to its rapidity in cooling, and vice versa.* It is, therefore, manifest, that this property militates against it as an agent in heating horticultural buildings, as it is well known to be an all-important point, in warming these structures, to obtain an equilibrium of heat for the greatest length of time, and with the least possible amount of attention, and experience has fully concluded that this is most effectually and most easily obtained by the circulation of hot water through wooden and metallic radiators and conductors.

Suppose, for instance, that a house, containing 4000 cubic feet of air, is required to be heated, from 32° to 60°, and suppose the external thermometer to remain stationary at this point; then, by calculation, we find that it requires double the amount of fuel to heat the atmosphere through the 28 degrees between these two points, by means of water, that it does through the medium of air, i. e., by direct communication, in each case the calorific action being in pretty exact ratio to the combustion, and both acting under the most favorable circumstances. This would, at first view, decide us in favor of hot air as a means of heating, in preference to hot water; and the fact that the heat becomes more rapidly sensible by hot air, has induced many to come to a premature conclusion on this point.

Let us, however, take another view of the position here alluded to, and consider the two methods in regard to their permanency of heating power. We find also, by calculation, that while the temperature of the house is maintained at 60° for 3·25 hours by hot air, with the same amount of combustion the temperature of 60° is maintained for 10 *hours by hot water*, or three times the period that the equilibrium is maintained by hot air. The same experiment shows that 2 bushels of coal will warm an equal volume of air in a hot-house the same length of time that 5·067 bushels will warm by direct connexion of its particles with the source of heat.

Now, in a large house, or number of houses, this saving of fuel would, in a few years, amount to the difference of cost between the two apparatuses, keeping out of the question the saving of labor, the cleanliness and neatness of the one compared with the other. In regard to these numbers, we may remark, that the calculations of some experienced and intelligent gardeners, drawn from accurate observation, have made the difference between the two methods still greater, in regard to the consumption of fuel, — placing this position in still stronger light than by the calculation here given.

This remarkable difference in the retention of heat is owing to the following causes.

First. The power possessed by the water [as already explained, see "Comparison of Water and Steam"] of absorbing and retaining a large amount of heat, and giving it off gradually, as the atmosphere requires it.

Secondly. Owing to the body of metal with which the water is surrounded, which also absorbs and retains a large amount

of heat, and parts with it slowly to the air by which it is surrounded.

Water is a better conductor of heat than air. Every gardener
well knows how rapidly a wet mat, or any other wet substance,
will carry off the heat in a frosty night, if laid over a hot-bed,
or green-house. In fact, the temperature of a frame under such
covering will fall quicker than if fully exposed. Yet the case
is different if the mat be dry, because the apertures of the mat,
and also the space between it and the glass, are filled with air
at rest, — because the latter is a bad conductor of heat, and the
former a good conductor. In a tank of water in a hot-house,
the thermometer will indicate a temperature probably 10° above
the atmosphere, while, by plunging the hand in the water, it will
feel about 10° lower. This arises from the power possessed by
the water of conducting the heat from the hand immersed in it.
The effect in all these cases may appear different, but the principle of action is the same. Water conducts heat rapidly from
a body warmer than itself, and conveys it to a colder one.

Let a stream of air be forced through a tube 100 feet in
length, entering at the temperature of 150°; by the time it has
travelled, by its own specific gravity, to the end of the tube,
it will be reduced to the temperature of the external atmosphere.
A stream of water, under the same circumstances, will travel to
the end of the tube with a very slight diminution of its temperature, probably only a few degrees, and will have heated the
tube, if a good conductor, to nearly the same **temperature as
itself** during its passage.

SECTION IV.

1. *Size of Boilers, and surface necessary to be exposed to the fire.* — In adapting the boiler of a hot-water apparatus, it is not necessary, as in the case of steam boilers, to have its capacity exactly in proportion to the quantity of pipe that is attached to it. On the contrary, it is sometimes desirable to invert this order, and to attach a boiler of small capacity to a considerable length of pipe. We do not mean, however, in recommending a boiler of small *capacity*, to propose, also, that it should be of small superficies; for the efficiency of a boiler very much depends upon the quantity of surface exposed to the fire. The larger the surface exposed to the action of the calorific influence, the greater will be the economy of fuel, and, therefore, the greater will be the effect of the apparatus.

In proposing the adoption of boilers of small capacity, however, it is necessary to accompany the recommendation with a caution against running into extremes, for this error has been the cause of the inefficiency of apparatus in many instances. In some boilers, we have seen the space allowed for the water so very small that the boiler was thereby rendered completely useless.

Too small a quantity of water, and too large a surface exposed to the fire, give rise to various evils, among which are the deposition of neutral salts and alkaline earths by the water which evaporates, contracting the water-way, and impeding circulation; and also preventing the full action of the fire on the exposed surface of the boiler.

But perhaps the greatest evil arising from this state of things, is from the repulsion of heat by the metal of the boiler. The quantity of water it contains being so small, and the heat of the fire very intense upon it, a repulsion is caused between the iron and the water, and, consequently, the latter does not receive the full quantity of heat. The repulsion between heated metals and water has been ascertained to exist, even at low temperatures, being appreciably different at various temperatures below the boiling point of water. But as the temperature rises the repulsion increases with great rapidity; so that iron, when red-hot, completely repels water, scarcely communicating to it any heat, except, perhaps, when under considerable pressure.

It is obvious that the extent of surface exposed to the fire should be in proportion to the amount of water contained in the boiler and the pipes; and it is easy to estimate these relative proportions with sufficient accuracy, notwithstanding the various circumstances which modify the effect. Calculating the surface which a steam boiler exposes to the fire at 4 square feet for each cubic foot of water evaporated per hour, and calculating the latent heat of steam at 1000 degrees, we shall find that the same extent of boiler surface that would evaporate a cubic foot of water, of the temperature of 52°, into steam, of which the tension is equal to one atmosphere, would supply the requisite heat to 232 feet of pipe, 4 inches diameter, when its temperature is to be kept at 140 degrees above that of the surrounding air. The following proportions for the surface which a boiler for a hot-water apparatus ought to expose to the action of the fire, will be found useful.

Surface of boiler exposed to the fire.					4 inch pipe.		3 inch pipe.		2 inch pipe.
3½ square feet will heat					200 feet, or		266 feet, or		400 feet.
5½	"	"	"	"	300	"	400	"	600 "
7	"	"	"	"	400	"	533	"	800 "
8½	"	"	"	"	500	"	666	"	1000 "
12	"	"	"	"	700	"	933	"	1400 "
17	"	"	"	"	1000	"	1333	"	2000 "

A small apparatus ought, perhaps, to have rather more surface of boiler, in proportion to the length of pipe, than a larger one, as the fire is less intense, and acts with less advantage, than in large furnaces. It depends, however, upon a variety of circumstances, whether it will be expedient to increase the quantity of pipe, in proportion to the surface of the boiler, beyond what is here stated; for, although many causes tend to modify the effect, the above calculation will be found a good average proportion, under ordinary circumstances. The effect very much depends upon the quality of fuel, the force of draught, the construction of the furnace, &c., which, from what has been already said on these matters, will show that they will, in a great measure, influence the intensity of the heat received by the boiler. It is always safest, however, to work with a larger surface of boiler, at a moderate heat, than to keep the boiler working at the maximum of its power.

There is another cause, however, that will tend to modify the proportions which may be adopted. The data from which the calculation of the boiler surface is made assumes the difference to be 140° between the temperature of the pipe and the air with which it may be surrounded; the pipe, in this calculation, being 200°, and the air 60°. But if this difference of temperature be reduced, either by the air in the house being higher, or by the apparatus being worked below its maximum temperature, then, in either case, a given surface of boiler will suffice for a greater length of pipe. For, if the difference of temperature between the water and the air be only 120°, instead of 140°, the same surface of boiler will supply the requisite degree of heat to one sixth more pipe; and if the difference be only 100°, it will supply one third more pipe than the quantity stated in the table.

It will, therefore, frequently occur, in practice, that the quantity of pipe, in proportion to a given surface of boiler, may be considerably increased beyond the amount which is given in the preceding table; because, in forcing-houses, the temperature of the air may sometimes be above the number of degrees here given, and frequently the temperature of the water may be below 100°,—the pipe not being required to be worked at its full heat; and, therefore, in both these cases, a larger proportion of pipe may be worked by a given sized boiler.

In order to estimate the quantity of surface which is acted upon by the fire, an allowance must be made for the flues which circulate round the exterior of the boiler, (and all boilers should be so erected as to admit of the action of the heat round their sides.) Thus, suppose an arch boiler (Fig. 35) to be 30 inches

Fig. 35.

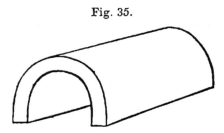

long, there will be about $8\frac{3}{4}$ square feet of surface exposed to the fire, that is, to its direct action underneath; and suppose, also, that there are four external flues, one on each side, — or supposing that the flue went all round the boiler, top and all, — we may calculate that nearly one half of the effect is produced by these flues which would have been obtained had the direct action of the fire been employed on a like extent of surface; therefore the flues will be equal to 5 square feet, making altogether $13\frac{3}{4}$ square feet as the available heating surface of a boiler of this shape and size, which we consider far superior to the old form of boiler, as shown in the following cut, (Fig. 36.) A boiler of the size

Fig. 36.

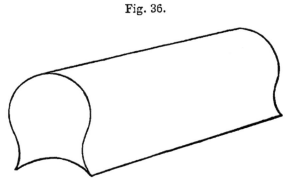

here described (Fig. 35) would be sufficient to heat about 800 feet of pipe, 4 inches diameter, when the excess of its temperature above that of the surrounding air is 140°, as before stated; a boiler of the same shape, 24 inches, has about 11 square feet of surface directly acted upon by the fire; one 36 inches long has $16\frac{1}{2}$ square feet of surface; and one 42 inches long has 19 square feet of surface; the increase being directly proportioned in the simple ratio to the length.

A circular boiler, 30 inches diameter, with a 9 inch circular flue running round the outside, will expose nearly the same extent of surface to the fire as the one just described, (Fig. 35,) both being the same length, and therefore the one will be as effective as the other; a slight diminution on the perpendicular length of the curve makes but little difference to its capacity for radiating caloric.

The surfaces of any size of this shaped boiler can easily be calculated by the same rule; but, instead of varying in the simple ratio of the length or diameter, it will be found to be proportional to the *square of the diameter*, so that the proportion of surface increases more rapidly than in the arched boiler. Thus, a circular boiler, 24 inches diameter, has $8\frac{3}{4}$ square feet of surface exposed to the fire; a 30 inch has $13\frac{3}{4}$ square feet; a 36 inch has $19\frac{3}{4}$ square feet; and a 42 inch has $26\frac{3}{4}$ square feet exposed to the fire; the small sizes having proportionally less surface, and the large sizes more than the high-arched boilers.

The rules which are here given regarding boilers, are framed to suit common occurrence, and intended to guide practical men who have the management and working of common hot-water apparatus. There are some cases, however, where apparatus of great magnitude is necessary, in which these rules will not apply without modification. But as such instances are comparatively rare, and, moreover, as no person that is a novice in the practical application of this principle of warming, will be likely to undertake, for his first essay, the responsible erection of an apparatus of great dimensions, it is the less necessary to enter at length into such cases as may be supposed to render any alterations of these principles necessary.

It may, however, be observed, that cases may occur where a peculiar construction of apparatus may be desirable; for instance, where, from a large quantity of required surface a furnace of very great power would be necessary; and, in that case, a boiler which exposes a large surface, while it possesses but a small capacity, would obviously be injudicious, because the intense heat acting on a small body of water would probably generate steam to a high degree of elasticity in the boiler, and not only produce much inconvenience, but neutralize the effects of what might otherwise be an efficient apparatus.

The nearer the rules here laid down for regulating the size of boilers are acted upon, the more efficient will be the working of the apparatus. There is no advantage whatever gained by using a larger boiler than is necessary to heat the pipes to their maximum temperature, — even though this temperature may never be required, — for, as the return-pipe should (if the apparatus be working right) bring in a fresh supply as rapidly as the flow-pipe takes it away, the boiler is . always kept full. It may be observed, that the circulation will be more rapid from a minimum boiler than from a maximum one, — that is, from a boiler whose capacity is rather below the proportion; while a boiler whose capacity is above the proportion of the pipes, has a slower circulation; and for all horticultural purposes, — though the former has some little advantage in the time of heating — the latter is decidedly to be preferred.

In the following section, (Sect. V.,) further information will be found on boilers, etc., where different methods of heating, in practical operation, are figured and fully described.

We may here state, in regard to the material for boilers for horticultural purposes, that cast-iron boilers, if properly made, will last much longer, and be also somewhat cheaper in the first instance, than malleable-iron ones, be the plates ever so good; the principle of durability resting on the former not being injured by oxydation so much as the latter. In both cases, however, the durability depends very much on the kind of water used; that least liable to form a deposition on the boiler being the best.

2. *Size and arrangement of hot-water pipes.* — Some controversy has arisen, among engineers, gardeners, and others, respecting the size of tube most suitable for the purposes of heating hot-houses. 2, 3, 4, 5, and 6 inch pipes have been used, and experiments instituted respecting the merits of each; from which it has been found that 4 inch pipes radiate more heat than any of the other sizes; and, consequently, the 4 inch pipes are now most generally used

The unequal rate of cooling of the various sizes of pipes, however, renders it necessary to consider the purpose to which they are applied. If it be desired that the heat shall be retained for a great many hours after the fire is extinguished, then pipes of larger dimensions must be used. Where a conservatory is very much exposed, and liable to fall below the minimum temperature during a cold night, then 5 inch pipe may be used, which will retain the heat longer than one of smaller size; but a double length of pipe should always be used in doubtful cases. But, as a general rule, no pipe should be used of more than 4 inches diameter, as the larger the pipe the greater the consumption of fuel, and more heat will be given out by 4 inch pipes, in proportion to the consumption of fuel, than by pipes of any other size.

The ordinary method of arranging hot-water pipes is by placing the furnace and boiler at one end of the house, and leading them along the front within a few inches of the wall. If the house be span-roofed, the pipes ought to travel completely round both sides; if single, or lean-to house, the pipe should pass along the front and return the same way; i. e., the flow and return pipes should be placed beside each other, as will be seen in the figures in the next section. The pipe ought never to run by the back wall of a house, except there be some reason to fear the entrance of frost in that quarter, which, in houses with thin walls, or those constructed with clapboards, is quite likely. In general cases, the heat rises with sufficient rapidity from the front, to prevent the entrance of frost at the back wall, unless it be near the bottom of the wall.

In general, hot-water apparatus is so constructed that when the smoke leaves the boiler, it passes immediately up the chimney, by which an incredible amount of heat is lost. I have seen the thermometer rise to 200° when placed at top of a chimney of this kind, and an amount of heat thereby lost nearly equal to the whole amount radiated in the atmosphere of the house. This is the case with many heating apparatuses, without the smallest notice being taken of the fact. On making this remark lately, to a most intelligent gardener, he doubted the fact of losing any heat by his chimney; while, on trying the thermome ter at the top of his chimney, we found it rise in a few minutes to 137°, after having travelled through 20 feet of flue through the back wall of the house.

Whatever apparatus be employed in heating a hot-house, the flue should always be taken advantage of. It must be remembered that smoke will not travel through a flue, — neither up nor down, — without first being rarefied by heat. The smoke, as already described, is, in fact, a body of gases emitted from the fuel by the action of heat, and a portion of this it takes along with it on leaving the furnace. In its passage, it communicates this heat to other bodies, as the flue; and more so, as the flue is in a position more or less horizontal. A flue, therefore, should, if possible, be carried the whole length before giving egress to the smoke, by which a great amount of fuel may be economized.

In laying down hot-water pipes, it is necessary to allow sufficient room for their elongation and expansion when they become hot. Want of attention to this has caused several accidents; for the expansive power of iron, when heated, is so great, that scarcely anything can withstand it. The linear expansion of cast-iron, by raising its temperature from 32° to 212°, is ·0011111, or about one nine hundredth part of its length, which is nearly equal to $1\frac{3}{5}$ inches in 100 feet. Therefore, it is necessary to leave the pipes unconfined, so that they shall have freedom of motion lengthways; and, instead of confining, as has frequently been done, facilities should be provided for their free expansion, by laying them on small rollers, or pieces of rod-iron, between them and the bearers on which they rest; for the contraction on cooling is always equal to the expansion on heating, and unless they can readily return to their original position when they become cool, the joints are apt to become loose and leaky, as indeed all cast-iron pipes do, that are exposed to sudden extremes of temperature.

Every hot-water apparatus should be provided with a supply-cistern attached to the boiler, or the pipes; the pipe leading from the supply-cistern should flow either into the return-pipe, or into the boiler, near the bottom. In no case should it enter the flow-pipe, as it is more likely to emit vapor, and the steam

16*

that may sometimes be generated on the surface of the water in the flow-pipe, would find egress, unless the supply-pipe were bent in the shape of an ∽ to prevent it, which is a very good plan; and, as a small lead pipe of about 1½ inch bore is sufficient to supply a boiler of considerable size, the pipe can easily be bent in any shape to answer the purpose.

3. *Impediments to circulation, &c.* ·The power which produces the circulation of the water in the pipes is the specific gravities of the two bodies in the return and flow-pipes; whether this force acts on a pipe 100 feet in length, or on one only 5 feet in length, the result is precisely similar.

Now it is evident that if this unequal pressure is the *vis viva*, or motive power, which sets in motion the whole quantity of water in the apparatus, in order to ascertain the exact amount of this force, it is only necessary that we know the specific gravities of the two columns of water, and the difference will, of course, be the effective pressure, or motive power. This can be accurately determined when the respective temperatures of the water in the boiler and in the descending or returning pipe are known.

As this difference of temperature rarely exceeds a very few degrees in ordinary cases, the difference of the weight of the two columns must be very small. But, probably, the very trifling difference that exists between them, or, in other words, the extreme smallness of the motive power, is very imperfectly comprehended, and will, perhaps, be regarded with some surprise, when its amount is shown by exact computation.

In order to ascertain, without a long and troublesome calculation, what is the amount of motive power for any particular apparatus, the following table has been constructed. An apparatus is assumed to be at work, having the temperature in the descending pipe 170°, and the difference of pressure upon the return-pipe is calculated, supposing the water in the boiler to exceed this temperature, by from one to twenty degrees. This latter amount will exceed the difference that usually occurs in practice.

By referring to the annexed table, it will be found that when the difference between the temperature of the flowing and returning columns is 8 degrees, the difference in weight is 8·16 *grains* on each square inch of the section of the return-pipe, supposing the height of the boiler A (Fig. 36, B) to be 12 inches. This height, however, is only taken as a convenient standard from which to calculate; for, probably, the height may, in many instances, be more than this, though it will seldom be less.

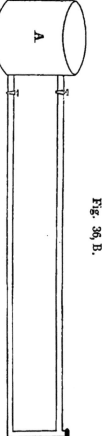

Fig. 36, B.

Now, suppose that, instead of 12, 18 inches was the distance between the two pipes, that is, between the top of the upper and the centre of the lower pipe, and the pipe 4 inches in diameter; if the difference of temperature between the water in the boiler and the return-pipe be 8 degrees, the pressure on the return-pipe will be 153 *grains*, or about one third part of an ounce; and this will constitute the whole amount of motive power of the apparatus, whatever be the length of pipe attached to it. If such an apparatus have 100 yards of pipe 4 inches in diameter, and the boiler contains, say, 30 gallons of water, there will be in all 190 gallons, or 1900 lbs. weight of water, kept in continual motion by a force equal only to *one third of an ounce.* This calculation of the motive power will vary under different circumstances; and, in all cases, the velocity of the circulation will vary simultaneously with it.

Difference in temp. of the two columns of water in degrees of Fah.'s scale.	Difference in weight of two columns of water contained in different pipes.					Difference of a column one foot high.
	1 in. diam.	2 in. diam.	3 in. diam.	4 in. diam.	5 in. diam.	per sq. inch.
	grs. weight	grs. weight	grs. weight	grs. weight	grs. weight	grs. weight
2°	1·5	6·3	14·3	25·4	33·6	2.028
4°	3·1	12·7	28·8	51·1	110·1	4·068
6°	4·7	19·1	43·3	76 7	211·7	6·108
8°	6·4	25·6	57·9	120·5	250·0	8·160
10°	8·0	32·0	72·3	128·1	317·5	10·200
12°	9·6	38·5	87·0	151·1	376·1	12·264
14°	11·2	45·0	101·7	180·1	390·9	14·328
16°	12·8	51·4	116·3	205·9	449·1	16 392
18°	14·4	57·9	131·0	231·9	522·0	18·456
20°	16·1	64·5	145·7	258·0	700·0	20·532

The above table has been calculated by the formula given with table IV., (see Appendix,) for ascertaining the specific gravity of water at different temperatures. The assumed tempera ture is from 170° to 190°.

It will be observed, in the foregoing table, that the amount of motive power increases with the size of the pipe; for instance, the power is four times as great in one of 4 inches diameter as in one of 2 inches, and nearly six times as great in one of 5 inches. The power, however, bears exactly the same relative proportion to the resistance, or weight of water to be put in motion, in all the sizes alike; for, although the motive power is four times as great in pipes of 4 inches as in those of 2 inches, the former contains four times as much water as the latter. The power and the resistance are, therefore, relatively the same.

These calculations are given with the view of showing how trifling a cause may impede the proper circulation of the hot water in pipes, and that, when once obstructed, how impossible it is for an apparatus to work. Trifling as this power may appear, yet upon its action depends entirely the efficiency of an apparatus. Seeing that the motive power is so small, it is not surprising that, by an injudicious arrangement of its parts, the motion may frequently be impeded and even destroyed; for the slower the circulation of the water, the more likely is it to be interrupted in its course.

There are two ways by which the motive power may be increased. One, to allow the water to cool a greater number of degrees between the time of its leaving the boiler and the period of its return through the descending pipe. The other, by increasing the vertical height of the ascending and descending columns. The effects produced by these two methods are precisely similar; for, by doubling the difference of temperature between the flow and return pipes, the same increase of power is obtained as by increasing the vertical height.

There are two methods of increasing the difference of temperature between the flowing and returning pipes. First, by increasing the quantity of the pipe, so as to allow the water to flow a greater distance before it returns to the boiler. Secondly, by diminishing the diameter of the pipe, so as to expose more surface in proportion to the quantity of water contained in it, and by this means to make it part with more heat in a given time.

The first of these methods, although the most practical, is necessarily limited, in some instances, to the length of the building to be heated, to which the length of pipe must be adjusted, in order to obtain the required temperature; and, as to the second, we have already enumerated many objections against the use of small pipes. Where the motive power, therefore, is not of sufficient strength, the increase of the height of the column ascending from the boiler must be depended on for an additional motive power.

In all cases, the rapidity of circulation is proportional to the motive power, and, in fact, it is the index and measure of its amount. For, if, while the resistance remains uniform, the motive power be increased in any manner, or in any degree, the rapidity of circulation will increase in a relative proportion.

Now, the motive power may be augmented, as we have seen, either by increasing the vertical height of the pipe, by reducing its diameter, or by increasing its length. If, by any of these means, the circulation be doubled in velocity, then, as the water will pass through the same length of pipe it did before, in one half the time, it will only lose half as much heat as in the former case, because the rate of cooling is not proportional to the distance *through which the water circulates, but to the time of transit*. If, then, by raising the pipes vertically, the difference between the temperature of the flow and return pipes be increased, it appears to be the most practical method of increasing the velocity of motion. The increased velocity, therefore, is indicative of increased power, and in a hot-water apparatus it is the velocity of circulation which enables it to overcome any extraordinary obstructions.

Neither the principle nor the practice of an apparatus is in the least affected by having an additional number of pipes leading out of, or into, the boiler; the effect is the same, whether there be more flows than return pipes, or, conversely, more return than flow pipes.

4. *Level of Pipes.* — Some persons have supposed that if the pipes be inclined so as to allow a gradual fall to the boiler in its return, additional power is gained. This appears very plausible, particularly with regard to some forms of apparatus, but the principle is entirely erroneous. This error appears to arise from treating the subject as a simple question of hydraulics, instead of a compound result of hydrodynamics. If the question were only as regards a fluid of uniform temperature, then the greatest effect would be obtained by using an inclined pipe; but the water in the pipes we are now treating of, is of varying density and temperature, which very materially alters the results.

Contrary to the ideas of some persons, the circulation of the water first takes place in the lower pipe; in consequence of the water in the boiler becoming lighter by the absorption of heat, the column of water in the return-pipe, being of greater density, forces its way into the boiler, when the water in the upper pipe falls into its place. Now, suppose the distance between the entrance of the return-pipe and that of the flow-pipe be 12 inches. This distance is neither increased nor diminished by any inclination of the return-pipe towards the boiler, the effective pressure being in both cases the same.

Discarding the erroneous hypothesis that the motion of the water commences in the upper pipe instead of the lower one, — and the motion commences at the entrance of the ower pipe into the boiler, which we have frequently proved, — it is, therefore, evident that there can be no advantage by making the pipe to incline from the horizontal level; for whether the water descends through a vertical or through an inclined tube, the force of gravity will only be equal to the perpendicular height; there must, therefore, be an equality of pressure on the boiler under all circumstances, whether the pipe entering the boiler be on a level, or inclined from its junction with the flow-pipe.

When it is necessary to sink the return-pipe below the level of the boiler, there must be a sufficient weight of water in the pipes, above its level, to overcome the perpendicular column that exists below the level of the boiler, otherwise the tendency of the lower column will be to a retrograde motion. The only way is to raise the pipe sufficiently to afford a perpendicular return-ing column of sufficient pressure to raise the water in the per-pendicular pipe attached to the boiler.

If the flow-pipe be carried on a horizontal level with the boiler, and the return-pipe carried *below* the level of the boiler, it is scarcely possible to obtain any circulation; and if this depth be much, no circulation at all can be obtained. We have seen some costly apparatuses completely useless on this account; and those erectors of heating apparatus, unacquainted with the principles of hydrodynamics, are very apt to commit similar mistakes. The velocity of circulation in such apparatus will be just in proportion to the difference of weight between the columns above and below the boiler.

It must not be supposed that water will not circulate in pipes below the level of the boiler; and much trouble and expense have frequently been incurred in consequence of being ignorant of this position. All that is necessary is to give the upper section of pipe a sufficient preponderance to raise the water in the lower one, allowing for the superior density of the water in the lower pipe. It, however, requires considerable judgment in adopting any such forms of apparatus as this, for many concurring cir-cumstances are essential to complete success. It should, there-fore, never be adopted when a common horizontal working apparatus can be introduced.

5. *Accumulation of air in pipes.* — It is necessary to make provision for the escape of air in the pipes, which sometimes so accumulates as to prevent circulation.   This is more especially the case when the apparatus is complicated, and has many turnings and vertical bends in the pipes.   It generally collects at the upper bends of the pipe, but this will depend very much upon the mode of supplying the apparatus with water.   It frequently requires the greatest care and the closest attention to discover where the air is likely to lodge, as the most trifling alteration in the position of the pipes will entirely alter the arrangements in regard to the air-vents.   Want of attention to this has been the cause of many failures, and the discovery of the places where the air accumulates is sometimes a matter of difficulty. For although it be true, in a general sense, that air will rise to the highest part of the apparatus, it will frequently be prevented from getting to the highest part by alterations in the level of the pipes, and by other causes.

As water, while boiling, always evolves air, it is not sufficient merely to discharge the air from the pipes on first filling them, because it always accumulates ; and, in many instances, it is desirable to have the air-vent self-acting, either by using a valve, or small open pipe ; but we have generally found a cock most convenient.

The size of the vent is not material, as a very small opening will be sufficient to allow the escape of air.   The rapidity of motion in fluids is inversely proportional to their specific gravities, as water is 827 times more dense than air ; an aperture which is sufficiently large to empty a pipe in 14 minutes, if it contained water, would empty it, if it contained air, in one second.   Air being so much lighter than water, it is of course necessary that the vents provided for its escape should be placed at the highest parts of the apparatus, for there it will always lodge when no impediment occurs to prevent it ; but it will sometimes be found necessary to have several in different parts of the apparatus.

Though it is perfectly easy to provide for the discharge of the air from the pipes, — as far as the mere mechanical operation is concerned, — it requires much consideration and careful study to direct the application of those mechanical means to the exact spot where they will be useful. We have frequently seen mechanics, who, though well acquainted with the practical details of the apparatus they were erecting, yet were perfectly ignorant of the principles on which it works; hence the success of such an apparatus must be entirely a matter of chance. Wherever alterations of the level occur, vents should be provided for the escape of air; and, as we have said, a small tap (or cock) will be the most convenient method of outlet.

In a complicated arrangement of hot-water apparatus, it is sometimes so very difficult to detect the various causes of interference, and the impediments which arise are often so apparently insignificant in their extent, that when ascertained they are frequently neglected. Those, however, who bear in mind how very small is the amount of motive power in any apparatus of this description, will not consider as unimportant any impediment, however small, which they may detect; moreover, they will immediately see the propriety of having the evil in question put right. But, in the more complicated forms of the apparatus, so many causes become operative in impeding the circulation, that the real cause of impediment may elude the detection of even an experienced practitioner.

We will now proceed to give a description, in detail, of various methods of heating, which come within the range of our own experience, accompanying the descriptions with sketches, by which their details will be more easily understood.

17

SECTION V.

THE heating of hot-houses, by any of the ordinary methods of warming these structures, has hitherto been attended with extravagant expense. The difficulty of obtaining, at a reasonable price, the means of keeping up the desired temperature, during long and severe winters, — the expense of the apparatus, — the annual cost of repairs, — the continual outlay for fuel, — together with the incidental expenses and trouble of working them, has, in many instances, proved a barrier to their erection, and has induced many to abandon the attempt, who had well nigh carried it into execution. Many lovers of exotic gardening have thus been diverted from the enjoyment of this pleasant and healthful pursuit; and hence it is of the utmost importance, especially to amateurs and others having small establishments, and who do not keep a regular gardener, that the internal arrangements of a plant-house, and, above all, the heating arrangements, should be so constructed as to be dependent upon the very smallest possible amount of time and attention, and likely to produce the least injury by neglect.

Among the numerous systems of heating lately applied to horticultural buildings in England, is one called Polmaise, from its having originated at a place in Scotland of that name, — the seat of the late Mr. Murray, near Sterling. The principles upon which this method is founded are not new, and the system itself, in other modifications, dates from a period much more remote than any other with which we are acquainted. This system is applied, in a more practical and perfect form, to the warming of many public and private buildings in this country. The very general adoption, however, of this system, does not, in **the** smallest degree, give us a warrant against its defects. It has been ascertained that air heated to a temperature of 300 degrees, becomes so deprived of its organic matter, and otherwise changed in its properties, as to be unfit for the sustenance of either animal or vegetable life, in a state of healthy and vigorous development, for any length of time ; and hence the admission of a current of highly heated air into a dwelling room, or into a well glazed hot-house, if no means are taken to restore its original properties, must, in a short time, become sensibly injurious to the animals and vegetables that are compelled to breathe it.

And this we find to be practically the case.   Every gardener, on entering a hot-house so heated, is immediately sensible of the presence of contaminating gases in the atmosphere, whether arising from the combustion of fuel, or otherwise, and he is too well acquainted with its effects on vegetative beings to allow his tender plants to absorb it; hence he takes immediate measures of modifying what he cannot possibly prevent.   It can scarcely be doubted, that a vast amount of sickness and diseases of the respiratory organs is, in a great measure, attributable to the same circumstance, especially in people of sedentary habits, who confine themselves to close chambers, warmed by currents of hot air, or highly heated stoves.   The latter, in this respect, is probably worse than the former ; for, in the one, the supply of air to be heated is drawn from the external atmosphere, and, consequently, is less likely to contaminate the air of the room, although, when conducted into the room at high temperatures, the atmosphere of the latter, without egress as well as ingress of air, must ultimately become so.   In the case of stoves, however, it is different, for by them the same atmosphere is heated over and over again, by convection.   The particles of air in contact with the stove first become heated, these expand with the heat, and, consequently, becoming lighter, rise, and the colder particles supply their place, which also expand, rise, and are in their turn replaced by others.   Here the supply of air to be warmed is drawn directly from the room itself; thus compelling the inmates to inhale the same contaminated atmosphere for days together, without mixture or admission of fresh air, except the small portion that finds an unwelcome entrance by the occasional opening of the door; and in the severe weather of our winters, with the thermometer below zero, this portion is frequently small indeed. The pleasure and ability of exercising our physical functions in cold weather, will be in exact proportion to the frequency of practice; and it is truly surprising, that with so much positive proof of direct injury resulting from continued confinement over highly heated stoves, many will, nevertheless, persist in so pernicious a custom, — a custom which is truly national, and which renders the influence of these stoves as baneful as that of the Upas tree, and sends thousands annually to an untimely and premature grave.

I have observed, by some articles that have lately appeared in an excellent horticultural periodical, (Downing's Horticulturist,) that this much talked of system of warming horticultural structures with hot air, called Polmaise, has been adopted by some individuals in this country. These individuals have been misled by the extravagant statements, or rather *mis*-statements, that have from time to time appeared in the Gardener's Chronicle, (of England,) by its talented editor and others under his influence. Those who have been in the habit of reading that paper in this country, and noticed the laudatory articles that have so frequently appeared in it, in favor of this method, yet unacquainted with the practical opposition it has received by numbers of experienced men, in every way qualified to decide upon its merits, can scarcely be blamed for adopting a system said to possess so many advantages over all others; and when it is considered that the gardening journal, which represents the opinions of practical men in that country, is but little read in America, — in fact, I may say, almost unknown, save by a few individuals, — it is not surprising that they should have been betrayed into the system supported by such authority. It is difficult, indeed, to account for the strong-headed and one-sided policy of the advocates and promoters of Polmaise. The fact is well known, that the system, and the defects connected with it, were thoroughly established many years before it was applied at the place from which it takes its name. In many places it had been tried, and found inferior, and far more fickle than the common smoke flue.* It originated at Polmaise Gardens, from the following circumstances : — A church in the neighborhood of that place had been warmed by a hot-air furnace, similar to those used in dwelling-houses in this country. A gardener at that place examined it, and thought it a good plan to warm his hot-houses ; accordingly, he applied something of the same kind to heat his vinery. The thing was entirely new to the worthy gardener, as well as to his employer, who sent an account of it to Dr. Lindley, of the Gardener's Chronicle, who forthwith espoused the system, extolled it to the skies, and induced various individuals to adopt it ; and those who would not, he straightway denounced as interested and dishonest men. The gardening community arose in arms, and waged war against their theoretical foes, until its so-called originators were confounded at the amount of opposition excited. No controversy connected with gardening was ever carried on with so much virulence as this one on Polmaise heating ; and no system has been so severely tested, to

---

* The premature encomiums so liberally lavished upon this system, by the zeal of its promoters, have neither shamed imposture nor reclaimed credulity. Deceptions seldom stand long against accurate experiments, and the mere charm of novelty soon vanishes, when economy and utility are both against it. The desire of notoriety, if nothing else, has too often induced parties to impose on the credulity of those who have not science enough to investigate its principles, nor practice enough to discover its defects. Nothing can more plainly show the necessity of doing something, and the difficulty of finding something to do, to obtain these paltry ends, than the getting up of this method of heating hot-houses ; and this, too, by those who know, or ought to know, better, and who ought to have rejected it with contempt. When a system has no intrinsic value, it must necessarily owe its attractions to theoretical embellishment, and catch at all advantages which the art of writing can supply. Trifles always require exuberance of ornament ; the building which has no strength or utility, can be valued only for the novelty of its character, or the money which it cost. It is certain that the advocate of a new system is less satisfied by its failure, than its success, even when no part of its failure can be imputed to himself, and when the fruits of his labor are tested by those who can discover their real worth. No man has a right, in things admitting of gradation, to throw the whole odium upon his opponents, and totally to exclude investigation and inquiry, by a haughty consciousness of his own excellence.

17*

prove its worth. Gardeners, amateurs, and all, entered the arena of experiment and discussion. Still its promoters would not flinch from their original position, and, right or wrong, would cram it down gardeners' throats, whether it was digestible or not; and that, too, without one tittle of evidence in favor of it, except ripe grapes in September, — a period when grapes would ripen themselves, without any artificial heat at all. Yet its cheapness and simplicity were its recommendation, and for some successive winters many went to work Polmaising their hot-houses, tearing down their furnaces, flues, &c., and converting them into Polmaise stoves, hot-air drains, and other appurtenances of Polmaise; but, after a short trial, and a good deal of plant-killing, they one and all abandoned the system with disgust. Still, amidst all this dust and dirt, and smoke and gas, created by the cracking of plates and the breaking of tiles, the Doctor maintained his ground, until, like the conquered hero, he was left alone in his glory, in the midst of the wreck and ruin he had created. What seems very strange, he never erected one, or caused one to be erected, at the Horticultural Society's garden, where he had unlimited control, and ample opportunity of so doing; and those who erected them by his recommendation and advice, were obliged to acknowledge them unqualified failures, notwithstanding all their alterations and improvements upon the original plan, which was simply this: — A hot-air furnace is placed behind the back wall, about the centre of the house; immediately opposite the stove there is an aperture in the wall, for the admission of the heated air into the house; directly in front and above this aperture, a woollen cloth is suspended, which is kept constantly moist by a number of worsted skeins depending from a small gutter, fixed on a frame of wood, which supports both the gutter and the cloth, the lower end of the latter reaching the ground. The cloth is made thicker in the middle, in order to equalize the heat, — an arrangement which is absolutely necessary; for if the cloth was an equal thickness all over, the centre of the house would be heated to a scorching degree, (by the rush of hot air,) while the ends would be comparatively cold. By means of drains under the floor, the fire-place is supplied with air from

Fig. 37.

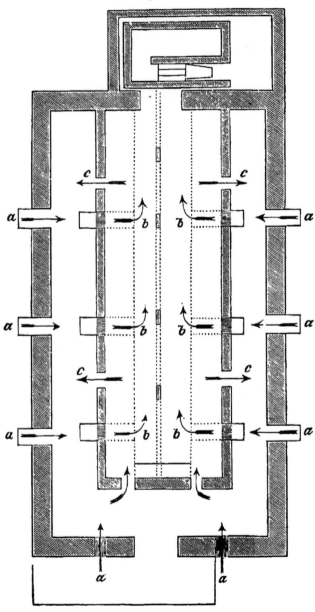

inside the house, part of which is used for the combustion of fuel ; the rest passes over the heated stove and enters the house through the apertures above noticed.

Fig. 38.

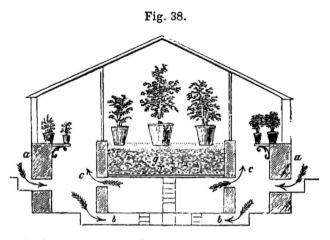

Such is the original system of Polmaise heating, which has created so much sensation in England, but which is now abandoned for some one or other of the many improved methods to which it gave rise, the most perfect and scientific of which, I have represented in the accompanying cuts, Figs. 37, 38, and 39. The arrows marked a, in the three figures, show the entrance of the cold air from the external atmosphere ; and its passage to the fire-place, beneath the floor of the house, is further shown by the arrows b, in Figs. 37, 38, and 39. Its passage over the hot plate, through the chamber, under the bed, and thence into the house, is marked by c, attached to each arrow in the three figures ; d, the fire-place ; e, a tank containing water, immediately over the cast-iron plate ; f, a small funnel, or tube, for supplying water to the tank ; g, (Fig. 38,) shows the bed on which the plants are placed, resting on cross-bars, and filled with pieces of brick, having a layer of sand or sawdust on top ; this can be converted into a stage, if desired. This is Mr. Meek's modification of Polmaise, from whom the drawing appeared in the Gardener's Chronicle, and was there represented as something very near perfection in heating, if not perfection itself. The above sketch is somewhat altered and simplified in the formation of the drains; and yet, in all conscience, it is complex and compound enough for a heating apparatus, as any person can see by a glance at the above sketches. It is difficult to discover wherein lies its superiority over the old smoke-flue, and it is clearly evident, that it has neither cheapness, simplicity, nor economy in fuel, to recommend it; and, as to its working, it is infinitely more precarious than the common flue, and the loss of heat is certainly much greater. This loss has been stated, by those who have tested its merits, to be at least one fourth of its whole heating power. Mr. Ayres, one of the most enlightened gardeners in England, stated, in a paper on that subject, published in the Gardener's Journal of 1847, that Mr. Meek wasted more heat from his one house, than he (Mr. Ayres) did from one fire that had *nine* different arrangements to work; and in a Polmaise apparatus that Mr. Ayres had erected, the waste of heat was enormous; that in ten minutes after the fire was lighted, he could ignite a piece of paper at the top of the chimney with the greatest ease; and when the same gentleman asked one of its strongest advocates the following question, "If you had a range of houses to heat in the best possible manner, would you abandon hot water for Polmaise?" he was answered, "No, certainly not."

I have quoted the opinion of Mr. Ayres, because he is well known to be one of the best authorities on matters of practical

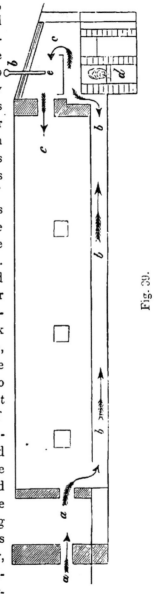

Fig. 39.

importance, connected with horticulture, at the present day, and his opinions are endorsed by almost every gardener of note in England. Mr. Fleming, of Trentham, and Mr. Paxton, of Chatsworth, as well as many others, regarded it as a thing utterly unworthy of notice. Mr. Ayres, in the same paper already quoted, puts to the advocates of Polmaise the following conclusive and unanswerable query. If Dr. Lindley, or any other of its advocates, can point to one place where the apparatus is at work, and as efficacious as a hot-water apparatus; if they can refer us to any one place, where we can see better productions than what have resulted from the use of hot water, why, says he, I am ready to spend five sovereigns to go to see it, and be convinced of my error in opposing it; but until then, it is mere nonsense to suppose that any responsible person will adopt it.

As an example of a combination of hot water and hot air, applied in a practical and scientific manner, the following system is superior to any other with which I am acquainted, especially for small houses. It supplies heat, moisture and air, either singly or combined. It consists of a cast or plate iron boiler, *a*, for containing the water; in shape it is not unlike a pretty large inverted flower-pot, with a hollow between its sides, about four or five inches wide, having one pipe entering near the top for the flow, and another at the bottom for the return, with a tube entering quite through to the fire-chamber, as represented at *b*, *c*, and *d*; then there is a hot-air chamber round the boiler and fire-place, as shown at *c*, *e*, *c*, Figs. A, B, and C; the boiler rests on a circular course of bricks, forming the furnace *f*, *f*, Figs. B and C. The whole is enclosed by the hot-air chamber, from which the air is conducted into the house, at *h*, and is supplied with cold air, both for the combustion of fuel, and drawing off the heated air, at *i*, *i*, Fig. C. The fire is fed through the door in the chamber, *j*, opposite which is a smaller door in the furnace, at *k*. In Fig. C is shown the door of the ash-pit, *l*, through which the ashes are drawn. We know of no apparatus, where a small green-house or conservatory is required to be heated, that will do it so effectually and economically as this. No particle of heat generated is lost, and in its simplicity is everything that a novice could desire. Here is nothing more

Fig. 40.

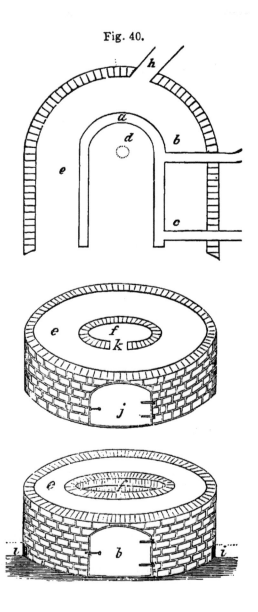

than a cone of cast or plate iron, with hollow sides, one hole for a flow, and one for a return pipe, (these pipes can branch into several directions, if necessary, on leaving the boiler,) and a channel through it, with a flange, or neck, on which to fix the smoke pipe ; build the boiler, thus formed, on a fire-place, with just distance sufficient below the edge of the cone for a door, to supply fuel ; this door should be quite narrow, in order to let the edge of the boiler as far down as possible. The hot-air chamber should be built of brick, and, if exposed to the atmosphere, should be at least one foot thick. In fact, the thicker the wall of the hot-air chamber is made, the better will the heat be retained. A tank of water is placed over the hot-air entrance, inside the house, for evaporation. If this system be not bungled in the construction, it will be found as cheap as any other, and the expenditure for fuel is but trifling. The circulation of the water is complete, and the air in the chamber is neither roasted nor burned, as it is chiefly received through the boiler, and, consequently, is possessed of more natural purity, which is so essential to vegetable life ; and it requires so little attention that any amateur can manage it without much trouble. Even in pretty severe weather, when set fairly agoing in the evening, it wants no more attention till morning ; set it right in the morning, and you may safely leave it again till night. Nor is it liable to accident or derangement. Not the least of its recommendations is its economy of fuel, — a circumstance of considerable importance, especially where the cost of fuel is high ; and, therefore, the economy thereof is of double moment to the proprietor.

We have never seen this system applied to large structures, but we have no doubt, were the apparatus made in proportion to its work, it would answer as well in large as in small houses ; at all events, there is no reason why furnaces and boilers of every description should not be chambered round in a similar way ; a very great amount of heat, that is now lost, would be turned to advantage, and I think it is not too much to say, that hot-houses could be heated at one half the expenditure of fuel.

The system of heating two, three, or more, houses with one boiler, is one of those valuable improvements which science, combined with mechanical ingenuity has devised, and which has been carried out in practice with the most gratifying success, — so much so, that in some places, separate apparatuses have been torn down, and this system adopted instead, merely on account of the fuel economized thereby. Among the many systems brought before the public, under the fine-sounding name of *improved*, it is doubtful whether any of them have given so entire satisfaction as the above, where it has been properly constructed. The facility so admirably afforded by this method of heating any of the connected houses in the space of a few minutes after it is found necessary, is certainly a great recommendation in its favor. In short, you have only to turn a tap, and the thing is accomplished. Fig. 41 represents the ground plan of four houses heated in this way, and most efficiently.

It will be seen from the plan, that the two end houses on the front are heated by the pipes flowing and returning into the pipes which supply the hot water for the two houses standing on the back. This is easily accomplished by having a tap on each pipe where it enters the house, so that either house may be heated, or both together, if required.

In the extensive forcing-establishment of Mr. Wilmot, at Isleworth, near London, no less than seven ranges of houses, each ninety feet in length, are heated by one boiler, and all are heated effectually, and that too for the purpose of forcing grape-vines. In many other places, in England, we know that this method has been adopted with the very best results.

In the plan here given, the box, (Fig. 42,) which is given on a larger scale, is situated immediately over the boiler. It may, however, be on the same level, or nearly so, and situated in any corner out of the way. The boiler here used is a common saddle boiler, and with a large apparatus, is probably the best boiler for general purposes. The apartments, *g g*, in the cut (Fig. 41) are offices for the garden, tool-house, potting-room, fruit-room, &c., and may be used as a mushroom-house. As the hot-water pipes pass through them, they are kept slightly warmed, and may be made useful as store-rooms and other kinds of garden offices

In some places in England, no less than eight or ten different

18

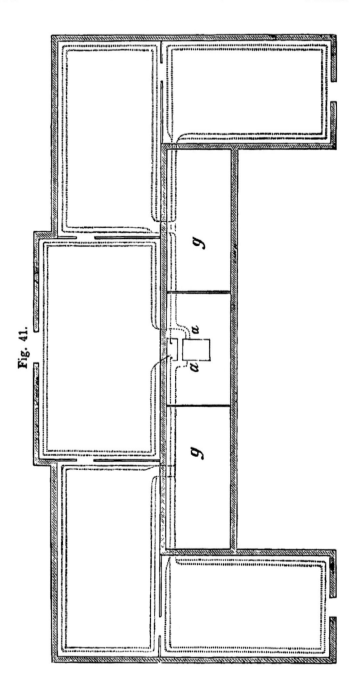

Fig. 41.

Fig. 42.

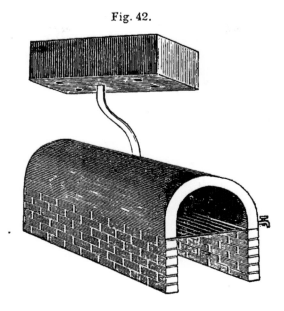

departments are heated by one boiler; some of them going at one time, and some at another, and sometimes all going together, and each having abundance of heat.* The convenience of this system cannot be too highly appreciated, especially when there are a number of small plant-houses situated near each other. For instance, suppose the boiler to be at work for one of the houses, which may be a plant-stove or forcing-house; well, you

Fig. B.

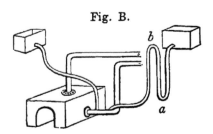

* Fig. B shows the common method of placing supply-cisterns. They may be placed in some convenient situation and attached by a small pipe to the apparatus. To prevent the escape of vapor, it is desirable to bend the pipe into the form shown at *a b*, as the water in the part of the inverted syphon at *a*, will remain quite cold.

go out before bed-time, and find the sky clear and frosty, contrary to your anticipations in the early part of the evening, — and how often do we find this really to be the case, — you enter into your green-house, and you find the thermometer travelling down rather quickly towards the freezing-point. Kindling fires is generally an unpleasant business at this time of night, and we are pretty often inclined to let the plants take their chance, rather than be at the trouble of doing it, even if it should cost us half a night's sleep through anxiety. Here, this unpleasant business is dispensed with, and the anxiety too, as well as the sitting up till the house is heated and safe for the night. You go to the tank or box, which is generally situated so as to be easily got at, in a recess made in the wall, perhaps, or immediately over the boiler, as represented in Fig. A; but, in any case, it should be so arranged as to be always of easy access from the houses. The arrangement of the pipes makes no difference, providing the accumulating tank be sufficiently elevated. The moment the water is put on, the circulation commences; in flows a delightful stream of hot water, warming the pipes as it proceeds through the flow and return; a vivifying glow of warmth pervades the chilly atmosphere of your green-house, and you can retire to rest without being troubled with anxious thoughts about your plants, let the weather turn as it may.

It may appear, that, by this arrangement, a larger quantity of fuel will be required for a single house, than if that house had an apparatus for itself. Not so, however; for, by close observation, it is found that the consumption of fuel is pretty nearly in proportion to the water heated, and that the heat given off by the pipes is in direct ratio to the heat absorbed by the boiler from the fire. Thus, if one house only be at work, there is only the water of one arrangement to be heated; and, consequently, only one return of cold water into the boiler, the rest being shut off. Now, if the water be *shut* off into the box, that is, the mouths of the flow-pipes stopped, there is no circulation; hence, there is no return of the cold water into the boiler, and, consequently, no absorption of caloric or combustion of fuel. Of course, more fuel is required to heat the four houses, than would be required to heat one, for the reasons stated, that the larger the body of cold water flowing into the boiler, and the larger the body of warm flowing from it, the more heat is carried away; hence, the more specific caloric is required, and the more combustion of fuel to produce it. But the proportion of fuel consumed to the proportion of heat generated by the pipes is found to decrease as the radiating surface is increased. This decrease amounts to nearly one third; for it is found that eight separate houses, or departments of a house, can be heated by the same quantity of fuel which it formerly required to heat five. This calculation was supplied to me by an intelligent gardener, of extensive experience, who made it from strict investigation into the working of the system under his own charge; and the statement is corroborated by the fact, that no case has occurred, to my knowledge, among many with which I am acquainted, and have examined, that has failed to give satisfaction.

This system has not the complex character which some have assigned to it, and which, at first sight, it would appear to possess; and, as to its cheapness, I believe little can be said about it, when placed in comparison with other hot-water apparatuses. I have had no means of calculating the difference, if any, between this apparatus and as many single ones as it may be substituted for. But it certainly appears, that four houses heated with one boiler and one furnace, would be cheaper than four houses heated with four distinct boilers and furnaces, the quantity of piping in both cases being equal; for then, three boilers and furnaces, or the cost of them, would be saved. This difference, however, will depend very much upon the distance the pipes must travel before entering the different houses. When the houses are situated close to each other, the difference must be very considerable. Some apparatuses of this kind have no box attached to them, and work directly to and from the boiler. I consider the box, however, as a very important appendage; not only because it affords greater facility for working the apparatus, but because any of the other arrangements may be repaired more easily, and parts may even be taken away without in the least affecting the working of the rest.

As I have already stated, pipes, in reality, radiate a very dry heat; though many think otherwise, because the air of a hot-

18*

house, so heated, is generally less arid than one heated by a hot-air stove. This arises from the fact, that, by hot-water pipes, a much larger radiating surface is presented to the atmosphere of the house than by any other method, and the heat is radiated at a lower temperature, and more equally diffused; hence, less moisture is carried upwards by currents of heated air and deposited on the glass by condensation. Thus, it is clear, that the larger the heating surface that is acted upon by the air, and the lower the temperature of that surface, the less moisture will be drawn from the plants and the atmosphere of the house. It is always desirable, however, to provide against aridity in the atmosphere, as heated air will have its supply of moisture, come from where it will; and if it cannot draw it from anywhere else, it will draw it from the plants, or whatever can supply the largest quantity under its influence. For this purpose, a number of troughs are made to fit on the pipes, made of zinc or galvanized tin. These troughs may likewise be made of earthen ware, and perhaps more cheaply than of zinc, though more liable to be broken. They may be filled with a syphon from the pipes, or by a common water-pot. When moisture is required in the house, an agreeable evaporation will be given off, and which can be rendered still more healthful, by putting in a few bits of carbonate of ammonia among the water, or common pigeon's dung, or guano. As the water warms, ammonia will be evolved into the atmosphere and greedily absorbed by the plants.

In recommending this system to the notice of those who may be entering upon the erection of hot-houses, we would state that we recommend it not only upon our own experience, but also upon that of others, whom we consider much better qualified to decide upon its merits. Nor do we mean to assert that it is the *ne plus ultra* of a heating apparatus, although, under certain circumstances, it is the nearest approach to it that has yet come under our observation. In making this statement, we do not wish to dispute the judgment of those who think differently, and who have opposed it more from a feeling of groundless distrust, than from any fact they can bring to bear against it. We have conversed with many who would prefer heating each house with its own fire and boiler, and be at the additional trouble of attending them too. This, however, springs from a fear entirely without foundation, and we are convinced that a little experience in the working of this double system of heating would so prove it. It is a very singular attachment which some people have for old methods and customs, that they will unflinchingly adhere to them, however little merit they may have to recommend them. Some individuals, with a self-sufficiency altogether incompatible with knowledge, will smile or sneer at what they are pleased to call the folly of enthusiasm, and, without seeming to be in any way sensible of the importance of whatever tends to the improvement of horticulture, regard these innovations merely as idle speculations of men who have nothing else to do but invent them; and while we cannot guard too much against the adoption of methods that will prove inconvenient in practice, although supported by theory, it is an injury to gardening, as an art, to give an unqualified opposition to systems that have proved their superiority, and are still capable of great improvement. This plan is not introduced under the deceptive cognomen of cheapness. Its cost will very much depend upon the circumstance of position, and may, after all, be much less than some of the costly and cumbrous apparatuses that are now in use. The easiness with which it is worked adds an additional item to its worth, for, when once set agoing, and understood, the veriest novice could manage it. *

* It is the common fate of new systems connected with the art of horticulture, that they are eulogized beyond their real merits by their advocates, and decried as strongly by their opponents; for every new system has always both friends and foes, each of whom are unwilling to adhere to the naked truth, and equally incapable of appreciating its merits with exactness. When a person invents, or fancies he has invented, something new, he is too much inclined to set a high value upon it; for, if it has cost him much labor, he is unwilling to think he has been diligent in vain. He, therefore, magnifies what is merely an alteration into an improvement, and probably prevails upon the imagination of others to fall into a false approbation of the system, and to regard that as a valuable desideratum which, at the best, was only a novelty. If durability and economy in working be allowed to constitute any part of excellence in a system, then this one has especial claims to our notice; a fact which cannot be said of many others.

Fig. 43.

Fig. 43 shows a house wherein provision is made for increasing the heating surface, when more than a very moderate degree of heat is required from the pipes. A box or tank, made of wood or zinc, is placed under the stage, and passes all round it on a level with the pipes. This tank is supplied by a branch pipe a, proceeding from the flow-pipe, and is provided with a tap at b, for shutting off and on the water when necessary. This is a most convenient arrangement; for, if a moderate heat only be required from the apparatus, then the pipes will be sufficient, and a very small fire will be required to heat them, as the quantity of water is small. When it is found necessary to increase the temperature of the house, the pipes being then tolerably warm, the water from the flow-pipe is admitted into the tank by opening the tap. The heat of the pipes is slightly reduced, but the radiating surface is increased, and the temperature of the house rises by an equal distribution of heat. It might be supposed that a quantity of specific heat is lost to the atmosphere, by drawing it from the pipes and throwing it into a body of cold water. Not so, however, as a little consideration will make sufficiently clear. Thus, if the atmosphere of the house be at 45 degrees, the water in the tank will be at 45 degrees also. Now, suppose the tank and the pipes to contain equal bodies of water, then, if the pipes communicate a portion of their heat to the tank, the temperature of the water in the tank will rise just as much as the water in the pipes will fall; for, if two equal bodies of water, at different temperatures, are mingled together, the temperature produced by the mixture will be the mean of their previous temperature. Suppose, for instance, that the temperature of the water in the pipes was 300, and that in the tank 45 degrees, and that, by the opening of the tap, the hot water in the pipes, and the cold water in the tank, were intimately mingled together, then the temperature of both would be 122·5 degrees. The temperature of both has been equalized, but the atmosphere of the house has lost none by the change, but rather gained, as the tank being 77 degrees above the temperature of the atmosphere, more heat will be diffused than with the pipes alone at double the temperature, and the object will be gained, namely, that of preventing the plants from being subjected to a high temperature, that are situated in the vicinity of the pipes. In the drawing, (Fig. 43,) the flow and return pipes are placed together, the other side being heated by the flue from the fire. This arrangement is intended to economize heat, and to save the expense of pipes, which, in some places, might be an object of importance, and even if they were not so, the plan is decidedly good. As for the tank, it is an admirable contrivance. Not only is the evil of having highly heated pipes for weeks and months together, directly under the roots of plants, prevented, but when the tank is once heated, a more agreeable and healthy warmth will be produced, and the equilibrium of temperature be maintained for a much longer time.

The tanks used may either be wooden or metallic. The latter are preferable, both on account of durability and radiation of heat, although wooden ones are much cheaper, and answer the purpose perfectly. Wooden tanks, if the wood be kyanized, or otherwise treated with a metallic solution, will last for many years, and produce a very agreeable warmth. Galvanized iron and zinc are now in common use for this purpose. The durability imparted to it by the process of galvanization, which prevents oxidation, is evident from the number of articles made of this material and exposed to the atmosphere. For horticultural purposes, this article is likely to become exceedingly useful; as every one is aware of the injury which ordinary hot-water pipes, and other metallic substances used in horticultural erections, are liable to from rust. Tanks made of this material give out their heat much more rapidly. But it must be considered that the same circumstances that would render them more quickly effectual, would also render their effect more transient. For pits and very small houses, the pipes and tanks might be made of this material. Its cheapness and lightness are important advantages in its favor; for, when heavy-cast metal pipes are conveyed to a great distance, the cost of carriage will nearly amount to the same sum as would purchase galvanized iron or zinc tanks, and convey them too.

In using this kind of tanks, the utmost care ought to be taken in supplying them with water. They ought never to be over one third or half full; the less water that is in them the better, compatible with safety. To work well, the water ought never to boil into them, otherwise the force of the steam will expand the metal, and, if the heat continue, it is liable to burst

Fig. 44 represents a house heated with a tank made of galvanized zinc. Next to the boiler there is a short piece of cast-iron pipe which prevents the zinc from being affected by the immediate action of the fire. The house from which this sketch was taken has been in use for some years, and has given perfect satisfaction, while the original cost was very small. The principal objection to the use of this material for heating purposes, is, as I have already stated, the rapidity with which it is heated, and the rapidity with which it parts again with its heat. This circumstance renders it a good conductor, but a bad retainer, of heat; useful where speedy and immediate action is required, but useless where a slow and long-continued radiation is necessary at a very low temperature, as, for instance, for bottom-heat, for propagating-beds, and for plant-stoves. In such circumstances, we should decidedly prefer wood, particularly for the first-mentioned purposes. In green-houses, and even in forcing-houses, it may answer well; for it must be admitted that the source of heat must ever be looked for at the boiler, not in the material of which the tanks or pipes are made. And, although the advantage of employing a material that will absorb the heat given off from the source to any extent, and part with it gradually, must be apparent, at least, when it is an object to take advantage of the heat so absorbed, store it up, so to speak, with the view of employing it when the action of the apparatus becomes enfeebled. The law by which this is effected is the same as that by which the two bodies of water become equalized in temperature by admixture as described on page 210. This is the law of equalization, which constantly tends to bring all bodies to an equal temperature. If, for instance, the hot water from a boiler be admitted into two separate tanks, one of wood and the other of zinc, then, by placing the palm of the hand upon the wooden tank, it will feel agreeably warm, while the zinc or tin one would be quite unbearable, if not burn, and this while the temperature of the water in both tanks was the same. The

Fig. 44.

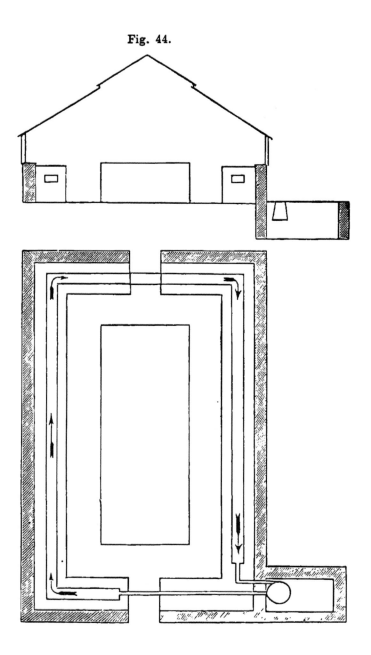

reason is, the metal is a good conductor, and quickly conveys away the heat from the water by imparting it to a colder body, while the wood is a bad conductor, and retains it. Again, the metallic tank will have the same temperature as the water within, before the wooden one is sensibly warm. In fact, the wooden tank retains and accumulates the heat, while the metallic one gives it off as soon as it receives it. None of this heat, however, is lost to the atmosphere of the house; for though the wooden tank parts with its accumulated heat more slowly, it as certainly parts with it, in the course of time, as the metallic one. It parts with its heat gradually till it is reduced to the same temperature as the atmosphere around it. A house heated with a wooden tank will maintain an average temperature with less expenditure of fuel than a thin metallic one, the other circumstances being equal, which is accounted for by the fact that when a house is suddenly heated, the warm air is forced rapidly upward, and, coming in contact with the glass, is rapidly cooled, descends, and is again warmed, till the warming surface is entirely deprived of its heat; then the temperature falls. On the other hand, when the heat is disseminated at a low temperature, the atmosphere is less agitated, and the ascending air less rapid in its motion. Not so much escapes through the laps of the glass, or is cooled down by the external cold upon its surface; and hence, the same quantity of specific heat maintains a given temperature for a longer time, when gradually given off, than when suddenly given off at a high temperature.

Although the sudden rise and fall of temperature by thin metallic tanks be apparent, we do not condemn their use for all purposes. As we have already said, they may be profitably used in many kinds of erections, and for various purposes; and I consider them worthy of more extended trials. But I do not believe that they will ever supersede cast metal pipes for the general purposes of heating by hot water; and for a retention-tank, I would decidedly prefer wood. Fig. 45 represents a house with a wooden tank, in which the water circulates by various divisions, after it enters from the flow-pipe. This tank was erected in a plant-house beneath the stage, as shown in the end section, Fig. 46, which may be objectionable as regards

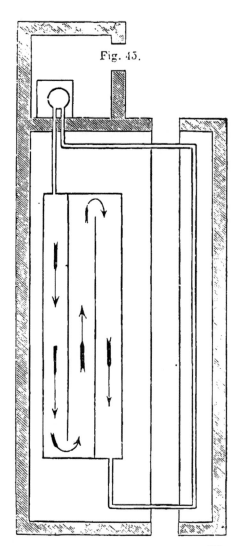

Fig. 45.

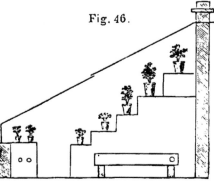

Fig. 46.

its position, but the chief object was to hold and retain a supply of warm water, which it did admirably, and effectually warmed the house besides. The manner in which this tank retained the heat after the fire had ceased to burn, impressed me with the idea that heat could be drawn off from the regular apparatus, and applied afterwards when necessary, or for any other purpose. I believe the heat generated by wooden tanks to be most favorable to the structural development of plants, as containing more moisture than heat radiated from either iron or brick, because the temperature is lower.

We have already remarked that the tank system of heating hot-houses has but very lately been brought into general notice, and still receives much less attention than its utility, simplicity, and economy claim for it; and where it has been used, it is chiefly as a medium of bottom-heat, for which it is undoubtedly superior to anything that has yet been applied. The efficiency of tanks in supplying atmospheric heat has been doubted by some and denied by others, without, however, as far as I can learn, bringing any practical facts to bear upon the subject. I am convinced that the system, rightly applied, will prove the doubts to be entirely without foundation. Simplicity in any system of heating is a point of incalculable importance; and when economy and adaptibility are combined with it, a claim is presented which facts only can overthrow. It is very true that we practicals are, many of us at least, prone to adhere bigotedly to any method with which we are acquainted, and which we have already proved safe and simple, and are unwilling to believe that any other method can be safer and simpler than itself. Gardeners are proverbially a cautious and thoughtful class of men; perhaps seldom directly opposing principles founded upon theoretical deductions, but frequently slow in instituting experiments with the view of establishing their truth. In these days of invention and progress, it is the duty of every one engaged in horticultural pursuits, and particularly gardeners, not only to make themselves acquainted with the views and opinions of other persons, but to test, by various counter-experiments, the conclusions they have drawn. No man is justified in regarding his knowledge of any system well-founded, except upon experiments and observations of his own.

Gardeners are, of all others, the best qualified to decide upon the merits of any system of heating hot-houses, and to ascertain its effects upon vegetable life. They are, by necessity, familiar with the habits of plants; and, by an instinctive practical knowledge, (if nothing more,) they are less likely to be deceived by the peculiarities produced by heat and cold, dryness and moisture, either in deficiency or in excess. The gardener is able to tell whether his plants be in vigorous health, or the reverse; whether they are suffering from atmospheric impurity, aridity, or stagnation; and, besides, the necessities of culture compel him to study the causes of such changes and conditions. All gardeners are aware that causes the most dissimilar will produce results in every way identical, while the self-same causes, repeated with the greatest care, and under circumstances where it was apparently impossible for them to be at variance with the first, will nevertheless produce results totally different; and the universal axiom, that like causes produce like results, would sometimes appear to be set at naught.

It has ever been a desideratum, as regards the heating apparatus, especially the hot-water kind, that there should be among gardeners a perfect knowledge of their details, and of the manner of repairing them. It is true we know when they become warm, and when they cool; but, as for the rest, once erected and the workmen gone, they are like a watch, or a doctor's prescription, — they may go wrong, and become unworkable, but we cannot put them right, nor scarcely discover what is the matter with them, till we send for the tradesman; and then, after an hour or two pulling and hammering, dusting and besmearing our plants, turning everything in the house topsy-turvy, lo! we are told that a joint had cracked, a collar had split, or some such mishap had befallen our apparatus. Facts of this kind will be in the experience of every one who has had much to do with heating apparatus. Now what I would urge is, that no part of a heating apparatus should be under ground, or buried in brickwork so far as it is concerned with the interior of the house Not an inch of it ought to be covered up with anything. It

ought to be all exposed, as far as possible, and of easy access.
Moreover, let it be simple in its arrangements; the simplicity
of any system is a plea in its favor. Some people, however.
despise simplicity, and would baptize everything about them with
confusion and complexity. We have met with people who fancy
that their green-house could not be heated without an array of
pipes, winding here and there, as intricately arranged as the
wheels of a watch, and as useless for the heating of their house
as the pillars that support the portico of their dwelling. One
would think that they admired cast metal pipes more than their
flowering plants.

One of the chief commendations of hot water, as a heating
power, is the facility with which you can bring it in contact
with the atmosphere of the house ; however simple the manner
in which it may be applied, it is not the less effectual ; and however commendable in other respects the warming of hot-houses
by improved methods of hot air may be, the channels and
chambers, the numerous hot and cold air drains, the under-
ground building of brick-work, and the multifarious intricacies
of its arrangements, are sufficient to deter any person from the
erection of such a complicated affair. This will be apparent
from a glance of the drawing of Meek's improved method of a
hot-air heating-apparatus, given on page 197. Such a concern
may do very well on paper, but it will not do in practice. It
may answer admirably as a plaything for amateurs, who have a
fancy for it, and nothing else to do with their time but to amuse
themselves with the motions of air ; but, as a method of warming a hot-house, no sane person will adopt it, when he can have
the thing done by a simple tank and boiler at half the expense.
As a system, it is good for naught. No person who understands
it will adopt it ; and those who do not know it, but will have it,
let them try.

The tank system may justly be regarded a real improvement
in heating, whether for top or bottom ; and it is the simplest,
and perhaps the cheapest, that has yet been brought under public notice. The sketches we have given are probably not the best
that could be adopted. It is yet open to great improvement, and
it would be premature, at present, to hazard an opinion upon

19*

what may hereafter be effected by it, by the friction of one idea against another in the course of experience. For small pits, filled with young stock, it is invaluable, as a pipe may be carried from a boiler, heating other structures, into a pit near by; and, these being easily covered at night, a sufficiency of heat may thus be conducted into them to keep the plants safe during the winter, without much increase of fuel, or any withdrawal of heat from the structure for which the apparatus was originally constructed. Green-houses are generally too much crowded in winter, and the adoption of dry pits for the conservation of plants that are somewhat hardy in their nature, is not so common as it ought to be. Pits might be so arranged as to obtain the superfluous heating power from other houses. This, in some instances, has been done, and it is likely that more will ere long be done in the same way; for if the vast amount of fuel, consumed by the general methods of heating, could be economically applied, without waste, it is not exaggeration to say that at least one third could be saved.

The hygrometrical and ammoniacal condition of the atmosphere of hot-houses has not received that attention, in connection with heating, which the importance of the matter evidently demands. We have books enough teaching us the effects of certain volatile and subtile fluids upon vegetable life, and exhibiting a multitude of facts which no person will venture to dispute; yet, in this matter, we practicals have, in a great measure, been deaf to the teachings of science, and blind to the lessons of nature. Practically, or experimentally, we have made but little inquiry whether invigorating or contaminating gases abounded in our hot-houses. Now, nature is either a good or a bad teacher, just in proportion as our knowledge of her immutable laws is limited or comprehensive. When we confine plants in a case of glass, as in a green-house, if we give them soil to grow in and water to drink, we are apt to think they ought to be contented; and if they do not thrive well and prove productive, we call them ungrateful, or very difficult to rear. Now, we ought to consider that plants feed by their leaves as well as by their roots, and that the volume of air in which the leaves are expanded, requires to be as regularly moistened and manured, as the body of earth in which they grow. It may appear vague and visionary to talk of manuring the atmosphere of a hot-house, but the thing is in reality neither so vague nor yet so visionary as it seems; for here science comes to our aid, and not only defines the vagueness, but converts the vision into a practical reality. It proves to us, both the benefits of manuring the air, and the manner of doing it. We know that plants derive a large portion of their food from the atmosphere; and we know, also, that the arid atmosphere of a hot-house is not always charged to a proper degree with these life-giving gases. An impoverished atmosphere must have the same effect as an impoverished soil. This is a well-known fact, and requires no demonstration to prove it. We are well aware that many plants will grow luxuriantly for years, suspended in the air, providing they be kept in a condition calculated to sustain them; but deprive them of these gases, and they will die, — deprive the atmosphere of its humidity, and they will quickly cease to exist as living plants. These vegetables absorb carbonic acid, ammonia, and water, from the atmosphere, by their leaves, even more abundantly than by their roots. This is especially the case with plants cultivated in pots; their roots being circumscribed into a small space, the nourishment is speedily exhausted, and if the atmosphere be at the same time robbed of its gaseous elements by artificial heat, the plants must perish, if this deficiency is not supplied to them by artificial means.

We have seen, that plants, even of a ligneous nature, will grow, form lignin, and proteine compounds, while suspended in a moist warm atmosphere, much in the same manner as plants do when growing in the soil. The amount of mineral matter they contain is indeed very small, and may be derived from the dust continually floating in the atmosphere, which is dissolved as it falls upon the leaves, and is absorbed with the atmospheric fluids. Here, then, we have plants subsisting upon the ingredients of the atmosphere; and experiments seem to prove that all plants are nourished by the same substances, in variable proportions, the chief of which are carbonic acid, water, and ammonia.

Experience has already proved the beneficial effects of these substances as fertilizers, not only of the soil, but also of the atmosphere. Plants watered with a weak solution of the salts of ammonia (smelling salts) will, in a few days, show their invigorating effects; and plants grown in a hot-house, with the atmosphere impregnated with ammonia, will exhibit, in a manner equally as striking, its beneficial influence. Every gardener is aware that plants, growing in frames or pits heated with fermenting manure, will, under ordinary circumstances, evince a much greater degree of luxuriance than in any other situation. In fact, dung-beds are considered an antidote for nearly every disease that plants are heir to, and not without a well-grounded knowledge of their effects; and hence, when a gardener wishes to invigorate sickly plants, he straightway plunges them into a hot-bed, and if there be any vitality left in the plant, it seldom fails in pushing out vigorously. Now nothing is more obvious than the fact that neither the heat, nor the moisture alone, produced this result; for if the plant had been plunged in a hot-bed warmed with the combustion of fuel, in nine cases out of ten the result would have been the very reverse. In fact, it is found, by long experience, that neither heat nor moisture alone will compensate for the removal of a sickly plant from the congenial warmth of a well-prepared dung-bed. Now, the question which presents itself for solution, in regard to this mode of heating, is, What is the cause of this difference, and how can it be otherwise produced? If we consider the effects due to the gases already mentioned, to be fully established, we will find that the secret of all this lies in the stimulating gases of the manure, which constantly surround the plants when exposed to the mild heat of a dung-bed. The old, and now almost obsolete, plan of warming forcing-houses with accumulated masses of fermenting manure, is well known; and the luxuriance of vines, forced by this method, is as well known as the method itself. This luxuriance was produced by the ammoniacal and other gases evolved during the process of fermentation; and though this method of forcing has been entirely laid aside, on account of its unsightly appearance, and the inconvenience of keeping up a constant supply of well-prepared manure, still the merits it possessed, by its ammoniacal properties have not yet been secured in any other mode of heating.

Suppose, then, that we have already solved the first part of the problem, by attributing these results to the beneficial action of gases arising from fermenting manure ; let us consider how we can produce those gases, under other circumstances, *i. e.*, without the presence of manure.

Ammonia is the result of a combination of the two gases, hydrogen and nitrogen, and has been hitherto known to gardeners, and applied by them, chiefly in the state in which it exists, and is produced by the decomposition of animal and vegetable matter, as in the formation of dung-beds, from which we can perceive it escaping in an uncombined state into the atmosphere. It is easily distinguished from all other gases by its powerful, penetrating odor. It remains, however, but a short time in this state, as it is speedily absorbed by porous substances, and by living plants, and combines with other gases, forming compounds ; with carbonic acid, for instance, forming the carbonate of ammonia of the shops, from which it can readily be disengaged and evolved into the atmosphere of a hot-house. Ammonia, in the state of a carbonate, is exceedingly volatile, and when a small portion is mixed with water, and the temperature raised to about 112 degrees, a large quantity of ammonia is evolved. This will be still better effected by mixing a small quantity of potash, soda, or lime, with the water in which the ammonia has been absorbed. The salt which held the ammonia in combination is taken up by these alkalies, and the ammonia, being exceedingly volatile, escapes into the atmosphere.

By dissolving the sulphate or carbonate of ammonia in hot-water tanks, or in thin troughs placed over the pipes and flues, an atmosphere may be produced strikingly similar to that of a dung-bed, and capable of producing nearly similar effects. Dung-beds are probably the most natural methods of applying artificial heat to plants ; and it is yet doubtful if we shall ever be able to supersede them in their invigorating influence, although much may be done to modify the existing evils of arid and unwholesome atmosphere in hot-houses. The mixture of guano, pigeon's dung, and various other substances, gives off large quantities of ammonia in warm water, and may be used with advantage instead of its salts. Tanks afford an excellent means of effecting this purpose, and are, not only on the score of simplicity and economy, but, also, in an ammoniacal and hygrometrical point of view, the best methods of producing heat with which I am acquainted.

The principal kind of structures to which the tank system of heating has yet been applied, to any extent, in England, are what are termed forcing-pits; and in these it has been extensively used, with much success. In this department of forcing, it has proved one of the greatest improvements of modern times. In England it is used on a large scale, in the culture of pines, vines, melons, cucumbers, &c., during winter; and, although in this point of view, it may not be deemed of equal importance in this country, where early forced fruits and vegetables are less demanded, it is, nevertheless, calculated to be of immense value to horticulturists in general, and plant-growers in particular. There is little doubt, but, ere long, an increasing demand for early forced fruits and vegetables, fresh from the forcing-house, will stimulate enterprising individuals to the erection of those cheap and simple structures, which could scarcely fail of being a profitable investment. A given space, covered with a glass roof, and otherwise protected, requires a comparatively small amount of fuel to maintain a tolerable degree of warmth in the soil, much less than is generally supposed. It is not my purpose to enter, at present, into the details of this question, and I merely notice it in connection with the subject of heating. By many it may be regarded as a mere speculative theory, which it certainly is, yet I think it worthy of more serious consideration.

In many of the English nurseries, tanks are used for stimulating the growth of their young stock, and in many kinds the annual growths are indeed remarkable. We have seen camellias, one year from the graft, as strong and vigorous as plants three or four years old under the old method of culture. Almost all kinds of green-house plants are benefited by being kept in tank pits, and we are inclined to think, if tank pits were more generally used by the nursery-men of this country, they would have their plants easier got ready for market, and they would require much less time to do so than is generally the case.

For amateurs wishing to "try to grow many things," but who have little time or money to spare for such purposes, one of these pits is just the thing he requires. A small pit, of this nature, with a very little attention, would keep a considerable number of green-house plants over the winter, and enable him to preserve a plentiful stock of bedding-out plants, such as verbenas, petunias, calceolarias, heliotropiums, penstemons, and many other pretty little things for the decoration of the flower-garden in summer. How much more pleasant and profitable would it be, for lovers of flowers, to have a little pit erected in some snug corner of their garden, instead of losing all their roses in winter, and storing their drawing-room plants, — their oranges, their camellias, their gardenias, oleanders, &c., — into the cellar, from which, of necessity, they are frequently taken half dead. Such a pit as I allude to may, or may not, be made to comprehend a narrow pathway along the back, — this would certainly be the most convenient, — and this portion might be covered with boards or shingles. This path would greatly facilitate the operations of watering, &c. Whether such a pit ought to be sunk below the ground, or placed on a level with its surface, will depend altogether upon the nature of the situation. Thus, if the position be a dry one, or admits of being made so by drainage, it should, by all means, be sunk two or three feet below the surface. But if the situation be very damp, it would certainly be bad policy to sink it so much; for whatever advantage it would gain in the way of protection, would be more than counterbalanced by the dampness which would be unavoidable. A pit, sunk in a dry situation, requires less fuel, even in the severest winters, than people generally suppose; and if covered from the frost, and kept dry, many plants will live over winter without fire at all. Plants are very much like animals, in regard to warmth; when once accustomed to a high temperature, they must have it continually; but inure them to the cold of autumn, and they will do with less heat in winter. This is not saying that we can change the *nature* of plants, and make them to endure a lower temperature than they can possibly, under any circumstances, bear. But we know that plants may be brought into a condition to enable them to survive a much greater amount of cold than they could otherwise have endured, and this, too, apart from the application of artificial heat. When half-hardy plants are destroyed by frost, its effects are most frequently visible at the collar, or lower part of the stem, arising from the intense action of the cold at the surface of the ground, which, in combination with moisture, first contracts and then expands the principal sap-vessels of the plants.

The annexed cut represents a double range of plant-pits, heated by wooden tanks. These tanks are supplied from a small boiler, placed in the centre, between the two pits; *a*, end section, shows the end of the tank, which is about six inches deep, and divided into two compartments, by placing a slip of wood up the centre, leaving a space at each end, for the water to circulate round. The arrows show the course of the water in its progress round the tank; the flow and return pipes are represented by dotted lines. These tanks are merely shallow boxes of wood, occupying nearly the whole inner area of the pits, and resting on piers of brick, or posts of wood; rough pieces of wood are laid crossways over the tanks, and a layer of broken bricks, (or sawdust, if the pots are to be plunged, which is desirable,) which forms the bottom, or floor, of the pit.

It is truly surprising how very little fire is required to maintain a perceptible warmth in these pits; and the growth of plants or vegetables of any description is astonishing. In some nurseries these pits are kept continually at work. The lights are entirely thrown off them, and the tops thoroughly exposed to the air; this prevents them from being drawn up tender and etiolated, and while their roots are stimulated with an agreeable warmth, they have, nevertheless, all the strength and hardiness of plants grown in the open air.

For the growth of early melons and cucumbers these pits are admirably adapted; they are equally efficient, without having the disadvantages of dung-beds. Their neat and tidy appearance gives them a place beside the other hot-houses, (which is not the case with hot-beds of manure,) to none of which they yield, in point of utility or interest.

If there is any one branch of exotic horticulture that possesses more extended interest than another, it is, undoubtedly, the cul-

Fig. 47.

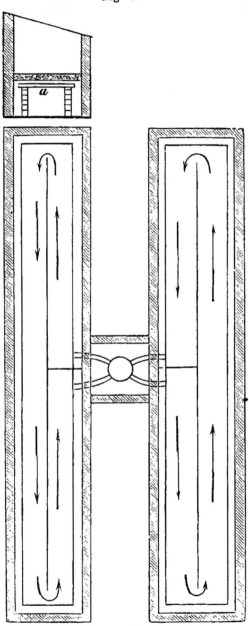

ture and early forcing of the grape-vine. The increasing importance of this branch of gardening will justify me in devoting a few pages to that subject. The culture of this fruit occupies a very high position in this country; volumes and pamphlets innumerable have been written about it, by practicals, theorists, and experimentalists, each one supposing he has discovered something, which, for want of more extended information, he calls "new," in the managing, heating, or ventilating of his vineries, when, lo! another starts up and knocks it on the head, and proposes a new nostrum; and every one is sure to find some ignorant enough to follow his advice. It might not be out of place here, to discuss some of the most important points which an extended experience has proved to be desirable, in the heating of structures for the culture of the vine.

And, first, let me remark, that nothing is more creditable than the use of the readiest and cheapest means at hand for securing a definite result. Whatever system may be thought of, it is desirable to understand the principles upon which it rests for its success. It must be borne in mind, in the outset, that no care in the culture of the vine, under glass, will compensate for a contaminated atmosphere, which should, at all times, approach to the natural summer purity and warmth.

Were we to analyze and bring into view the first principles of horticulture, and make ourselves masters of the various effects produced upon the grape-vine under glass, and the causes, we should often smile at the ludicrous importance we attach to particular methods of practice. A blind man, by habit, will often walk along a devious path, with quagmires and pitfalls on either side of him, and safely too, whether at midnight or noon-day; and we often follow the example of the blind. We, in one way or another, acquire the faculty of performing certain operations with a life-like certainty, though in the same degree of mental darkness as regards the power of deviating from the beaten track without committing egregious errors. In all such cases, there can be no doubt that it is wise to follow the old trodden path, till we can more plainly see which is the safest for our particular case. It does not so much signify which of the best methods we adopt, provided the science of culture has given us sufficient light upon the nature of the endless variety of means and methods that are before us, which, however defective they may seem in the hands of the inexperienced, may be safe and certain in the hands of the skilful practitioner.

It cannot be admitted, however, that in carrying out this principle it is unnecessary to scrutinize, with the utmost exactness, the facts for or against any particular system, which the fancy of gardeners or amateurs may choose to follow. The very fact that there are so many systems of warming hot-houses, gives increased force to the call for minute record of experiments. Upon no safer principle can our knowledge of horticulture be based, so that those who are its patrons and votaries may follow principles, founded upon facts, and not upon speculations. Hence they would not have to endure the inconvenience and risk of being dependants upon plausible theories, which practice may prove to be absurd.

A great deal that might be said on vineries, in regard to heating them, can have but a local application ; and, in some places, no application at all, inasmuch as the diversity of climate in the different states would render the erection of an apparatus at one place necessary, which would be absolute folly in another. The erection of a powerful and expensive heating-apparatus is only required where the forcing of the vine is desired in winter, under difficulties of intense cold and long-continued frost, as in New England. To these latter circumstances the following method will chiefly apply.

Figure 48 shows the plan of a winter vinery, i. e., one for forcing in winter; a is the border, underneath which is an arch of brick, forming a chamber, through which the hot-water pipes are made to travel, after going round the house inside for atmospheric heat. The cold water returns again into the boiler at b.

As far as I can learn — and I have made many inquiries — this system of applying heat, in connection with vine-growing, has not yet been adopted in this country; still, it may be in use, since the obvious utility of it must have been apparent to those who are engaged in the culture of hot-house grapes. To recommend such an expensive system as this, for all occasions

Fig. 48.

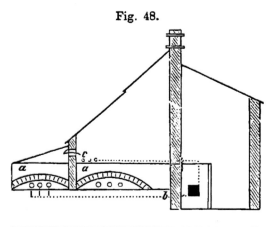

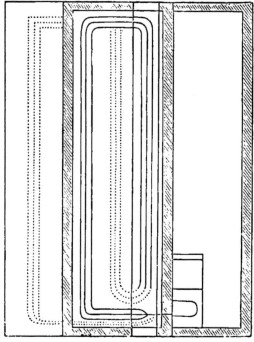

and under every circumstance, would be folly. But its utility in winter-forcing, especially where the soil is damp and naturally cold, will be obvious to any one of much experience in these matters. The greatest success has attended the application of border-heating, in England, where an enormous amount of money and labor is annually expended upon the forcing of grapes, and where they are produced in great perfection all the year round.

I have said that where early forcing is practised, and the soil and sub-soil of a cold, retentive nature, the adoption of some method similar to the above is almost indispensable to general success. I wish, however, to be rightly understood, and not to mislead, and therefore advert to what every gardener knows well, that good grapes are sometimes produced under the entire neglect of all the ordinary precautionary measures resorted to by good gardeners for the purpose of securing success.

In support of this method of heating borders, I will briefly advert to the opinions of some of the leading gardeners in England. Mr. Fleming, gardener to the Duke of Sutherland, at Trentham Hall, writes to the Gardener's Chronicle, four years ago, to the following effect: — " Shrivelling was common here, until the system of keeping up a bottom heat in the vine borders was introduced. Since then there has been no appearance of it, except in a late house last year. In the month of August we had a great deal of rain, which penetrated the border, and the weather was for a few days very cold, and the grapes, which up to that time were swelling beautifully, received a check, and shortly after many of the fruit-stalks shrivelled."

In the same paper Mr. F. makes the following statement, which is the strongest evidence of the utility of the system that has come under our notice: — " I am so convinced," says he, " of the advantage of this practice, that I would prefer the introduction of flues under every vine border about the place, did circumstances permit." This method is also employed at Welbeck, with the greatest success. There the soil and sub-soil are heavy, cold, and wet; and without some such precaution, grape-growing would be but a barren business. But by this method of chambering the borders, and other good manage

20*

ment beside, the most abundant crops are obtained. Mr. Roberts, of Raby Castle, author of a treatise on vine culture under glass, and a good authority on the subject, says : — "Fault has been found with me for recommending heat to the roots of vines by fermenting manure, on account of its unsightliness ; but practice convinces me that without a corresponding degree of temperature betwixt the root and top, you cannot produce good grapes. I intend, however, to do away with the unsightliness of manure, in my new vine borders, by heating them on another plan." Such is the testimony of men who stand first in their profession, — men of undoubted probity and extensive experience, and who, as authorities on these matters, may be fully relied on. No one, who once has seen the extensive gardens which they superintend, will dispute the propriety of the practice of placing fermenting manure on the surface of a vine border. But I must differ in my opinion from Mr. Roberts in regard to its effects. It may not be positively injurious, but Mr. Roberts has failed to prove that it is positively beneficial. Moreover, if he has succeeded in imparting a temperature to his vine border equal to the atmosphere at which he keeps his vinery, he must have a body of manure equal in bulk to the vinery itself. Heat travels with extreme slowness through the damp, confined air of dung-beds, and the difficulty of getting heat to travel downwards is well known. A body of fermenting material may communicate its heat to the mere surface of the soil on which it lies ; but the moisture it absorbs from the atmosphere, as well as its saturation by rains, is communicated to the soil in place of heat, so that in reality the good produced is nearly, if not altogether, counterbalanced by the evil. The plan Mr. R. intended to adopt has not, as far as I know, been made public ; but probably it was some kind of chambered border, with artificial heat radiating beneath it.

The annexed drawing represents a chambered border, heated with a hot-water tank, which is supplied with water from the pipes when it can be spared from the atmosphere of the house, by a tap fixed on the pipe, as shown at *a*, in the end section. If the water is allowed to flow into the tank from the boiler for the space of an hour, a sufficiency of heat will be communicated

Fig. 49.

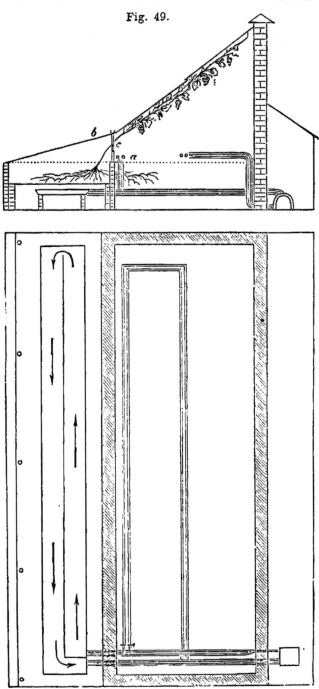

to the chamber, and the border for maintaining a perceptible warmth in the latter for twelve or fourteen hours. A division is made in the tank for the circulation and displacement of the water, as shown by the arrows in Fig. 49. Pigeon-hole walls are built across the border, about five feet distant from each other. Upon these rough pieces of timber are laid, as a bottom to the border; a layer of brush-wood (small branches) is laid over the timber to prevent the soil from falling through upon the tank. The rest should be filled, to the depth of two feet, with a good turfy material, with a plentiful admixture of whole bones and rough pieces of charcoal, to render the mass as porous as possible, for the admission of the heat upwards, as well as to maintain an equality in the moisture of the mass. Shutters are provided for covering the border, which may lie upon the same angle as the roof, or otherwise, as the front wall of the house corresponds to the curb in front of the border. Ventilators are placed in the front wall, beneath each light, for the admission of air into the house; and when air is required by these front ventilators, the shutters covering the border must be tilted at the lower side, when the air passes across the border, through the front, into the house. We consider this mode of arrangement for the border cheaper and better than that of arching the chamber, as shown in Fig. 48, although both are equally effectual, and may be adopted as circumstances may suggest.

If chambered borders be found so beneficial in England, for winter forcing, where the frost seldom penetrates more than a few inches into the ground, and rarely continues for more than a few days at a time, — a week or two, at the longest, — surely it must prove equally if not more serviceable in the New England states, where the winters are so intensely cold as to render the forcing of grape-vines at that time next to impossible. Still, if the forcing of this fruit can be carried on at mid-winter, at a reasonable cost, there is no reason to suppose that it would be unprofitable, even at the low prices at which grapes are usually sold in the principal markets of this country. All cultivators are aware that the profits of fruit culture are just in proportion to the economy with which good crops can be produced; and this is more especially the case in the culture of exotic fruits, there being more room for the exercise of skill in their production.

Suppose, for instance, that we take the calculations of Mr. Allen, in his treatise on the Culture of the Grape-vine, where, in pp. 69, 70 and 71, he estimates the quantity of fermenting manure, necessary for the covering and warming of a border 100 feet in length to cost $700; which, together with the other items of management, — repairs, fuel, interest on cost, etc., — to amount to $1120. The produce of a house so heated and managed, according to his calculation, is on an average 1067 pounds of fruit. I do not intend to dispute the accuracy of these calculations, although they appear startling enough. And doubtless Mr. Allen has had data sufficiently accurate and authoritative, from which to draw his deductions; and hence I consider myself justified in making them partially the data of mine.

And, admitting the beneficial effect of fermenting manure to be all that its advocates claim for it, let us compare the calculations above, with the cost and working of chambered borders; and, by balancing the two together, we shall be the better able to estimate the merits of each on the score of economy.

In order to effect this, I have been at some pains to obtain the probable expense of such a border as that represented on page 232, Fig. 49; and, in making my calculations, I have placed my figures rather above than under the estimate; so that, should I make any error, it will be on the most favorable side.

To make a chambered border 100 feet long, we have —

For brick work,	$200
Timber to form the bottom of the border,	60
Tank,	50
Extra piping for do.,	10
Extra fuel,	15
Excavating the border,	45
Shutters, &c., for covering do.,	100
	$480

Now, if we subtract 480 from 700, (the cost of manure,) we **have a** saving of $220, the very first season; or, in other words, the manure required for one year costs more than the making of this border, by $220. Then, if we estimate the annual expenditure on account of the border, for heating, repairs, etc., to be $25, we have $700, the cost of manure, minus $25, the cost of the tank border, which gives an annual saving of no less than $675 by this method of heating.

It may be supposed that a body thus situated over a hot-water tank, might be too rapidly dried by the ascending heat. But this is only a supposition; and in practice it amounts to nothing more, for the warmth generated by the tank is so gradual, and spread over so large a surface, that the heat is equally distributed, and no part of the mass is overheated, or one part heated above another. And, indeed, one would scarcely believe, from the small quantity of heat thus generated, that so striking an effect would be produced; of course, the border must not be allowed to get too dry. Nor will this be a matter of so much difficulty as may appear, as two or three good soakings with water, — or, what is better, weak liquid manure, — will generally suffice, until the weather permits you to uncover the border during the middle of a wet day, covering it up again before evening. The operation of watering will be much facilitated by having a hose fitted to the tap of a cistern containing rain-water inside the house; and no hot-house of any kind should be without such an appendage. If the mechanical texture of the soil be good, the water soon finds its way through. The larger portion of the moisture being held in suspension by the lower stratum of soil, becomes gradually warmed by the tank, and is again carried upwards by the heated air; so that the roots of the vines have the full advantage, not only of the heat, but of the moisture. The abstraction of heat may be in a great measure prevented, in excessively frosty weather, by laying a few inches thick of straw, or stable litter, immediately over the soil beneath the covering. This is merely a precautionary expedient, and, though useful, will seldom be necessary.

In the formation of a chambered border many alterations and improvements will suggest themselves to the mind of the practical man, which could not be very conveniently represented in the accompanying sketches. For instance, as a covering, instead of shutters, I would decidedly prefer glass; and where there are plenty of spare sashes about the place, they might be used in this way, with much advantage, just as spare sashes are used for covering peach and apricot wall-trees in England. But suppose that sashes of sufficient length were provided for the purpose; the expense would probably be counterbalanced by the advantage gained. For a house 100 feet long, 25 sashes would be required, which, at 3 dollars each, would be 75 dollars; a very trifling sum when a desirable object is to be attained by the judicious expenditure of it. And, in this case, although it may appear injudicious to some, the object is, in my opinion, sufficiently important to justify this expense. Light absorbed is productive of heat, especially if the absorbing body be of a dark color, for then it is absorbed without being again reflected upon the transparent medium. Hence we see the advantage of having the border covered with a body admitting light; and the soil of which the surface of the border is composed, of a dark color, that the heat which falls upon it may be absorbed and retained.

For winter forcing, small houses are decidedly preferable to large ones. Houses about 25 or 30 feet long are sufficiently large, and are more easily heated, and more convenient to manage. Even in the milder climate of England, small vineries are preferred to large ones, and are found to be more profitably worked. Above all things, loftiness should be guarded against, as being the very worst feature in a forcing-house, as the heated air continues to ascend upwards; and, unless the external atmosphere can be admitted at the top, the vines at that portion of the house will always be in a state of vegetable suffocation; a fact of too frequent occurrence, in lofty houses, even in summer, and which is rendered still more injurious by the present defective methods of ventilation.

A few words more regarding the permanency of these borders. Assuming that a proper command of heat, both for the atmosphere and the soil, is obtained, the question has been asked, How long will borders, so circumscribed, continue to supply a house of grape-vines with the requisite nourishment? This question has hitherto proved a drawback to the adoption of these borders by many who have, in every other respect, the highest opinion of the advantages to be derived from them.   In short, they are afraid the vines will exhaust the soil within the limits of the range allowed to the roots, and then fail in producing a crop for the want of food.   Now I think a very little consideration will prove this to be a groundless fear.   Supposing the soil to be the principal repository for the nutriment of the vines, and that it should contain all the substances in abundance, whether solid or gaseous, which form their structure and produce their fruit; yet it is not necessary to form this border into a mass of nitrogeneous matter to produce these results.   Plants, in this respect, are as bad as animals; and a vine-border may as readily be poisoned with excess, as impoverished for the want of proper elements of nutrition.   Now I maintain, and I do so upon experience, that the grand requisite to be looked to in the formation of a vine-border is its condition as regards texture, and not its chemical properties.   The first secured, the latter can be added, not only when it is first made up, but annually afterwards, and each subsequent time, with as much advantage as at the beginning.   The food of vines consists chiefly of the elements, carbon, hydrogen, nitrogen, and oxygen, in some state of combination, together with certain inorganic compounds, amounting to only about 7 per cent., as silica, salts of lime, magnesia, iron, potash, soda, and other bases, combined with sulphuric, phosphoric, carbonic, silicic, humic, and other acids.   These substances can be supplied, in a liquid state, in quantities more than sufficient for the actual requirements of the vine.   But their efficacy will very much depend upon the freeness, porosity, and other mechanical qualities of the soil, favorable to the decomposition and recombination of these elements.   The general method of renovating a vine-border is by incorporating about half its bulk of manure, to the manifest destruction of many of the best roots, — for the best are always on the surface, — besides incurring a vast amount of labor and expense, which labor and expense would be sufficient for at least a dozen years.

Salts of ammonia, for instance, in their various states of combination, are known to exercise a powerful influence on the growth of grape-vines.   Now, by adding, say, 10 tons of the best manure to the borders, we supply them with about 85 pounds of ammonia, in the form of sulphate, carbonate, nitrate, and muriate of ammonia. Now, the same quantity will be furnished by one quarter of a ton of good Peruvian guano, at probably one half the cost, while its application, in a liquid state, is more immediately beneficial to the vines. The same salts are supplied from urine, which ought to be collected in tanks for that purpose. By the addition of these elements, an impoverished border, incapable of yielding one fifth of a crop, has been enriched and made to produce good crops of fruit. As I have said, however, I would have a border made, say, 12 or 16 feet wide, of good open material, not over-rich in nitrogeneous matter. but abundantly mixed with lumps of charcoal, and plenty of bones ; a quantity of common lime-stone (carbonate of lime) might be laid on the bottom, and mixed through the mass. With a border so formed, about 2 feet deep, and 14 feet wide, by the regular application of nutritious elements in a liquid form, and proper management in other respects, the most abundant crops may be produced, for at least a quarter of a century.

We are well aware of the arguments that are brought to bear against shallow vine-borders in this country, from their greater liability to become dried up by the parching droughts of summer. But here this argument can have no application, as the season of forcing is at that period when the ground is saturated with wet, and little or no abstraction of it by the atmosphere. And as the temperature of any piece of ground is nearly in exact proportion to the amount of water it contains, so it follows that a vine-border saturated with water must necessarily be colder, and consequently more injurious to forcing plants, than a dry one, even without heat ! It is true, a border may be drained, and all superfluous and stagnant moisture carried off, but even the driest and most silicious soils have a certain capacity of suspending moisture in their pores, and as this capacity is greater in soils containing much organic matter than in those of a more sandy nature, it follows theoretically, — and we find it so in practice, — that rich borders are colder and wetter than the common garden soil. I believe this is a fact which no one will dispute. But however warm vine-borders may be by their natural position, or rendered so by artificial

21

drainage, they must still be very far from the internal temperature of the house, — a fact which requires no calculation to
prove it.   And hence, although the vine, of all other fruit-bearing plants, will accommodate itself to circumstances apparently
the most unpropitious, and will stand forcing in mid-winter better than any other fruit we can place into a hot-house, still, it
cannot be expected that we shall arrive at anything like perfection in its produce by winter-forcing, under the present methods
of cultivation.   And we know that, whatever can be said in
favor of carrion-borders, no mere aggregation of organic matter
will suffice for the production of grapes, especially in winter,
if the principle of life be impotent, and the functions of the plant
impaired, whether by natural or artificial causes ; and nothing
is more likely to weaken the one, or impair the other, than
placing the roots of vines in an ice-house, and the branches in
an oven.

In close connection with the foregoing subject is a system
which has engaged no inconsiderable share of attention in England, and may probably be employed with equal advantage in
this country.   The system to which I allude, is forcing by hot
walls covered with glass.   It has now become common to build
garden walls hollow, and heat them with hot water, with flues,
or both, and by covering them with temporary roofs, consisting
either of spare sashes on hand, or by having sashes made for
the purpose.   By this means, a range of portable houses may
be constructed upon any walls adapted for that purpose, at a
very inconsiderable expense, compared with that of permanent
houses. — (See Part I. *Construction of Walls.*)

Fig. 50 shows an end section of the wall ; *a a*, ties across the
wall, at regular distances, for the purpose of strengthening the
fabric ; *b*, the pipes for hot water, or the situation of the flue, if
that method of heating be adopted ; *c*, the furnace and boiler,
placed in a recess of the wall, as shown in the ground plan ; *d
d*, the returning pipes, or the position of the returning flue, if
pipes are not used ; *e*, the projecting support for the sashes
under the coping ; *f*, the lower supports for the sashes, consisting of timber posts driven into the ground, that no obstruction
may be presented to the roots of the vines by a brick wall ; *g*

Fig. 50.

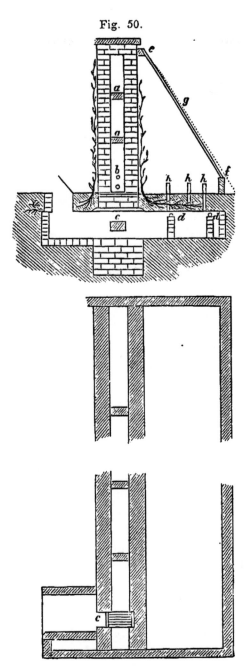

the sashes; *h h h*, are square tubes of wood, penetrating the soil to the chamber beneath, to let the heat rise up into the space confined within the sashes and the wall; the number of these openings depending entirely upon the weather, and the season of forcing. They may be closed with a lid when the sashes are removed. From the foregoing description, it will be perceived that an erection of this kind has all the advantages of a house, — at least as far as grape-growing is concerned, — without the consequent expense, and when once all the materials are properly adjusted, they can be removed, or replaced, by almost any gardener, without the aid of a tradesman. The rafters are merely fastened to a plate of wood, about one foot broad, and two inches thick, by means of iron pegs, as at *e*, in the end section, and also at the bottom to another plate, similar to the one above, and fitted into the posts at *f;* the sashes are fixed to the rafters by means of a latch, or thumb-screw, placed within reach of the operator, for the facility of admitting air. This is effected by letting down the sashes to any distance, and supporting them by notched brackets, or letting them down to the ground, if necessary, as shown by the dotted line, at *f*.

This method may be adopted without having any cavity beneath the border, and, of course, will be cheaper, although we would decidedly prefer such a cavity, did circumstances permit. The advantage of hollow walls, warmed by some method, has been long well known to gardeners, and so highly are they thought of in England, that scarcely any garden of consequence is without them. Indeed, in the majority of seasons, the culture of the vine, peach, nectarine, apricot, and fig, — even on walls, — would be a very precarious and uncertain business, although the method of covering such walls with portable glass has but very lately been brought into use, and, now that glass is cheaper in that country, is almost certain to be extensively applied to this purpose. In one or two cases we have seen this method adopted with astonishing success, and without any cavity, or any other preparation than the common border and wall of the garden. In one place we had forty feet of a wall thus covered with spare sashes; the space included some peach and fig trees, in excellent bearing condition, and well set with buds, giving promise of a fair crop. The wall was covered with the glass on the first of February, but no heat was applied to the wall until the beginning of March. The sashes were fixed exactly as we have described in the foregoing sketch. The wall and fire-place were precisely the same, but no cavity was beneath the border. The result was, that the crop ripened five weeks earlier than those on the same wall, uncovered, without heat, and nearly four weeks earlier than those on the same wall, with heat, and covered, in the usual way, with netting. Now this was merely an experimental result, without much previous preparation, save the covering up of the wall a month earlier than the warming commenced, — if, indeed, this can be called a preparation, — being an absolutely necessary prerequisite to success, under any method of forcing. When the warm weather set in, the sashes and rafters were taken away, and the enclosed part received, during the season, the same treatment as the other portions of the wall.

In forming a hollow wall, there will be quite as much saved by the internal cavity as will suffice to warm it, as only about one half the quantity of bricks are required ; and even without a heating apparatus, hollow walls are superior to solid ones, for horticultural purposes ; for, under all circumstances, they are found to be both warmer and drier. The addition of a heating apparatus, however, will render the wall a very useful auxiliary to the forcing-house, and the cost will be amply compensated by the utility. By looking at the foregoing plan, it will be seen that the furnace is placed in the foundation of the wall, with a few steps to descend to it, the whole being covered with a trap-door, leaving nothing unsightly open to the view of the visitor.

In many parts of this country, grapes are frequently overtaken by the autumn frosts, before they are ripened, and in many others, they do not ripen at all. Now, it is obvious, that it is neither owing to a deficiency of sun-light, nor a deficiency of heat, for in Britain the quantity of both are much less, and the quality of the latter less powerful for the maturation of fibre and fruit ; and yet it is common enough to have good crops of (what in America are called foreign) grapes, on the open walls. In ordinary seasons, the black Hamburg, Muscadine, and Fron-

21*

tignacs, ripen well in the open air, on southern aspects; and in all seasons they succeed in ripening their fruit, in tolerable perfection, on hollow walls, with a little heat in spring, if the season be backward, and a little in autumn, if the season be late, *i. e.*, if cold weather should set in unusually early, which it frequently does in Scotland; and yet we have seen tolerable crops of grapes produced north of the Tweed, on heated walls, without any glass at all. This statement may be received with incredulity by some, who have had poor success in the cultivation of foreign grapes, in the open air, in this country, under circumstances of climate unquestionably more favorable than can be found in any part of the British Islands. We believe this statement will be corroborated by the testimony of every one who is acquainted with the nature of the climate of both countries. In fact, so much are people in this country impressed with the unfavorable nature of an English summer, that in all journals, magazines, periodicals, and papers, of every description, we, without one single exception, find it qualified with the words, dull, gloomy, austere, wet, cold, damp, dripping, and many other appellations of similar import, which it is not my present purpose either to confirm or confute. But as there has not, as yet, been (as far as we can learn) any general cause assigned for the general failure here, there is but one inference that can be drawn from the above statements, viz., that there must be, in this country, something wrong, or something wanting, in the modes of cultivating foreign grapes, in the open air. It cannot be said that the summers are too hot for the grape-vine; for there is hardly another plant in the vegetable kingdom, that will bear a greater amount of natural or artificial heat, or greater alternations of heat and cold, under circumstances otherwise favorable. There is no degree of heat, to which natural vegetation is subjected in this country, under which it will not flourish, provided the intense rays of the noonday sun be not concentrated upon its foliage; and it is a well-known fact, that grape-vines will not produce fruit abundantly when they are not in a favorable aspect. There can be little doubt that we must look to the condition of the plant, during the spring and autumn, to enable us to reach the cause, and this appears to be substantially proved, by the invariable success that has attended the culture of foreign grapes in cold houses, as well as other facts suggested by experience, in the culture of that noble fruit.   The value and importance of the grape-vine have already induced me to dwell longer on this subject, in connection with heating, than I intended; I therefore consider it foreign to my subject to enlarge further on its culture, although there is great room for speculation, theory, experiment, and practical improvement.   Indeed, it would be difficult for the practical horticulturist to take hold of a subject affording a wider field for successful experiment, and holding out brighter hopes of beneficial results.

Before concluding this chapter on heating, I will briefly notice another system, more, however, on account of its novelty, than applicability to the warming of hot-houses, although it has, in some instances, been applied to this purpose.   I refer to the method of heating, by which the hot air is carried along by the power of a steam-engine.   This system is applied to the warming of large factories in England, and has been also applied, with apparent success, in some large nursery gardens, in Germany. The following description is from the pen of Mr. Marnock, the able editor of the "Gardener's Journal," (Eng.,) and drawn from his own observations of the apparatus, while visiting the gardens of Baron Hugel, near Schonbrunn, where the system was in operation at the time.

"The most remarkable feature about this garden is the mode of heating, which we shall now attempt to describe.   In the first place, there is a large fire-place constructed; through this fire-place two or more pipes are introduced; the pipes are of cast-iron; one end of these pipes communicates with the common atmosphere, the opposite end being introduced into a large box, or flue; in this flue is placed a fan, driven by a steam-engine, which fan is made to revolve in this air-flue, at a short distance from the fire-place.   It will readily be understood, that, when the fire is in action, with those iron pipes passing through it, and terminating in the large air-flue, the revolving action of the fan, in a direction to draw the common atmospheric air through the iron pipes in the fire-place, will also force the heated air onwards to the other end of the flue, and thence through tin, zinc, or any other kind of pipe placed there to convey it away. By these means it is conducted (*i. e.*, the heated air from the box) into the different stoves and green-houses. Each house, or, rather, each compartment, is provided with a supply-pipe and a tap, by which heated air is admitted by measure, and of course regulated according to the requirements of the plants. We could not clearly ascertain the exact size of the fire-place, but we saw some iron pipes, which we were told were similar to those in use in the fire-place for heating the air, and we supposed them to be about six inches in diameter. These pipes, as they are exposed to the action of a strong fire, become greatly heated, and the air, in passing through them, becomes intensely hot and dry, consequently, deprived of its oxygen and aqueous properties. Here, however, no evaporating pans are used for moistening the warm air, as in common hot-air furnaces, and the method adopted for supplying the heated air with moisture is quite as novel as the system itself. To effect this, a steam jet is played into the hot-air flue, immediately before it enters the different compartments, and Mr. Hooibrink, the gardener to Baron Hugel, stated that he admitted the steam according to the nature of the plants cultivated in each apartment. Thus, he allowed so many feet of steam for his orchards; so many for his stove plants, and so many for his common green-house plants; thus each kind of plants is supplied with steam, according as it requires a moist or dry atmosphere.

" Thus, if we are rightly understood, there is, first, a large fire-place; through this fire two or more cast-iron pipes, six inches in diameter, are passed; they are so placed as to be subjected to the most intense action of the caloric produced by combustion; one end of these pipes is exposed to the external atmosphere, the other ends enter a large oblong box, on a level with the pipes, in which is placed a fan, similar to those used in small fanning mills. This fan is made to revolve with considerable rapidity, by the power of a small steam-engine, drawing the atmospheric air inwards through the tubes exposed to the fire, and forcing it onwards through the main conductor, and thence into the smaller tubes leading to the right or left, up or down, as the case may be, for the supply of the mansion, and the several hot-houses, all of which are heated by the same apparatus. Here the air is moved and replaced, not only by its own density, as in the common methods of hot-air heating, but it is drawn rapidly inwards by the suction of the fan on the one side, and driven onward by its propulsive power on the other; and thus it appears the heat travels with great rapidity, and within a few minutes after the heated air is turned on the apartment, and moistened as may be desired by the jet of steam already described."*

* Fans are now frequently employed for effecting ventilation, and are generally connected with the heating apparatus. They have never, as far as we know, been employed for this purpose in any kind of horticultural structures, although we can see no reason why they should not be so. This will be again referred to, when we come to treat on that part of our subject.

PART III. VENTILATION.

SECTION I.

PRINCIPLES OF VENTILATION.

1. THE ventilation of hot-houses, either in summer or winter, constitutes an important, if not *the* most important, item of their general management, as it bears more directly upon the condition of the external and the internal atmosphere. It requires, therefore, the strictest attention of the gardener at all seasons of the year. For, important as other things connected with exotic horticulture may be, — light, for instance, — it is, nevertheless, more under the gardener's control, and more subject to his will. He places his plants within a transparent medium, which is, in its general surface, impermeable to the atmospheric air; and he forms for them an artificial atmosphere, which is invigorating, or the reverse, according to his knowledge of the laws that regulate the atmosphere, or the general principles of aerometry. Notwithstanding the many discoveries that have been made regarding the properties of air, I have been unable to find any work bringing these discoveries to bear upon the airing of hot-houses. It is true this must be accomplished by the "practical" man, and the sooner we begin to think about it the better. Every horticulturalist, no matter what his department in the vineyard may be, soon discovers the necessity of maintaining a continual warfare with its different conditions of purity and impurity, its aridity, and its moisture. We have, indeed, various theories propounded by physiologists, regarding the power of plants to withstand these vicissitudes, some of which have their general principles as yet enveloped in a mist of shadowy vagueness.

Many remarkable facts, however, might be mentioned relative

to the qualities and quantities of certain atmospheric elements which plants are capable of sustaining in deficiency or in excess. And the one or the other of these conditions appears to some of the species a natural and even a necessary circumstance. The degree in which vitality is sometimes retained by plants, under the most unfavorable conditions, for a period to which it is difficult to assign a limit, is one of the most interesting and curious circumstances in their economy. Instances have been related of the growth of bulbs, unrolled from among the bandages of Egyptian mummies. Although there is good reason to believe that deception has been practised on this point, upon the credulity of travellers, still there is nothing impossible in the asserted fact. Light, heat, and moisture are the cause of the development of these curious structures, and their forms become expanded under the additional agency of atmospheric air. Now, when removed from the influence of these, there is no reason why a bulb, if it can remain unchanged for ten years, should not do so for a hundred; and if for a hundred, why not for one thousand years? The vitality of seeds under similar circumstances appears quite unlimited. *

In the first chapter of this treatise, we have ventured to assert that light is of more importance to plants than air, although we are aware that this point is open to much discussion, from the fact of some plants being adapted to thrive under the almost total deprivation of it. These, however, will generally, if not solely, be found to consist of plants in the lowest orders of organization, such, for instance, as the algæ, some of which, possessing a bright green color, have been drawn up from the depth of more than one hundred fathoms, to which the sun's rays cannot penetrate in any appreciable proportion; and also the fungi, which have been found growing in caverns and mines to which no rays from the sun, either direct or reflected, would seem to have access. These facts, however, do not greatly affect the

---

* Melon seeds have been known to grow at the age of 40 years, kidney beans at 100, sensitive plant at 60, rye at 40, and there are now growing, in the garden of the Horticultural Society, raspberry plants raised from seeds 1600 or 1700 years old. — [*Lindley's Introduction to Botany.*]

accuracy of our assertion, for we find that all the highly developed organisms, such as we cultivate in our hot-houses, are only adapted to exist where they can be daily invigorated by the sun's rays. This fact is very strikingly illustrated in the effect produced on tropical plants growing in hot-houses in the northern latitudes, where, deprived of the intensity of the sun's rays, under which they naturally luxuriate, they seem completely changed by the long absence of the luminary on whose cheering influence they depend. In such cases, no quantity or quality of air will compensate for the loss of the sun's vivifying beams.

In the management of hot-house plants, the attentive observer cannot fail to perceive the remarkable effects produced upon certain kinds of plants by the circumstances in which they are placed, as to heat, light, and air; and hence the propriety of arranging plants in hot-houses, not merely according to their heights and colors, but also according to their habits and requirements in relation to these elements.* Some plants will endure an intensity of solar light, without injury, which would utterly paralyze and suspend the functions of others; some will luxuriate in an arid temperature, in which others would be destroyed, and some require daily supplies of fresh air, while others will exist even in a healthy state for years where the atmospheric air is, one would think, almost excluded. Even in nature there are many striking exemplifications of these facts. A hot spring in Manilla islands, which raises the thermometer to 187°, has plants flourishing in it and on its borders. In hot springs near a river of Louisiana, the temperature of which is from 122° to 146°, have been seen growing, not merely the lower and simpler plants, but shrubs and trees. In one of the Geysers of Iceland, which was hot enough to boil an

---

* For example, the common weeds, called chickweed, groundsel and Poa annua., evidently grow at a temperature very near that of 32°, while the nettles, and mallows, and other weeds around them, remain torpid. In like manner, while our native trees are suited to bear the low temperature of an English summer, and, in most cases, suffer if removed into a warmer country, such plants as the mango and coffee-tree, etc., inhabitants of tropical countries, soon perish, even in our warmest weather, if exposed to the open air. — [Lind. The. of Hort.]

egg in four minutes, a species of chara has been found growing and reproducing itself; and vegetation of an humble kind has been observed in the similar boiling springs of Arabia, and the Cape of Good Hope. One of the most remarkable facts on record, in reference to the power of vegetation to proceed under a high temperature, is related by Sir G. Staunton, in his account of Lord Macartney's Embassy to China. At the island of Amsterdam a spring was found, the mud of which was far hotter than boiling water, and gave birth to a species of liverwort. A large squill bulb, which it was wished to dry and preserve, has been known to push up its stalk and leaves, when buried in sand kept up to a temperature much exceeding that of boiling water.

Again, we have observed plants exceedingly tenacious of life under the deleterious influences of carbonic acid, sulphureous, chlorine and other gases. We have seen a number of different kinds of plants placed in a close frame, and fumigated with sulphureous gas, the greater part of which were destroyed, though a few of them were uninjured. This fact has been observed by many in the fumigating of their green-houses with tobacco, when some of the tender sorts would be sensibly injured by the smoke, while others, though receiving a much larger portion, bore it with impunity.

It is evident, however, that though many plants will live for a short time under these circumstances, a certain condition of the atmosphere, as well as a given amount of light and heat, is necessary to the performance of their functions, and the perfection of their flowers and fruit. The fact is well known, that if we take a healthy plant from the light and airy green-house, and place it in the room of a dwelling-house, it will become sickly, and ultimately languish and die; if it be placed in a dark, cold cellar, its death will be more speedily produced. In like manner, roses grown in a forcing-house in winter are less fragrant than those grown in the warm sunshine of summer. In general, plants grown in the summer months form secretions more active, in every respect, than the same kind of plants grown in a hot-house, under the clouded skies of winter; and even in our finest forced fruits and vegetables this is perceptibly the case.

22

2. Much discussion has taken place upon the question whether or not vegetation is, upon the whole, serviceable in purifying the atmosphere; that is, whether plants give out most carbonic acid or most oxygen. Priestley maintained that the latter was the *only* effect of vegetation, and that plants and animals are thus constantly effecting changes in the atmosphere which counterbalance one another. Subsequent experiments seem to show, however, that the carbonic acid given out during the night, equals or even exceeds in amount the oxygen given out by day. But this might be owing to the employment of plants which had become weak and unhealthy, by being kept in an impure atmosphere, previous to being experimented on, and which had not been exposed to a fair degree of light. Dr. Daubeny, of Oxford, has recently shown that, in fine weather, a plant, consisting chiefly of leaves and stems, if confined in a capacious vessel, and duly supplied with carbonic acid during sunshine, as fast as it removes it, will go on adding to the proportion of oxygen present so long as it continues healthy; the slight diminution of oxygen and increase of carbonic acid which take place during the night bearing no considerable proportion to the degree in which the contrary effect occurs during the day.*

Thus we see that the two great organized kingdoms of nature are made to coöperate in the execution of the same design, each ministering to the other, and preserving that due balance in the constitution of the atmosphere, which adapts it to the welfare and activity of every order of beings, and which is quickly destroyed when the operations of any of them become

---

* Plants decompose carbonic acid during the day, and form it again during the night, — the oxygen they inhale at that time entering again into combination with their carbon, — and, during the healthy state of a plant, the decomposition by day, and recomposition by night, of this gaseous matter, are perpetually going on. The quantity of carbonic acid decomposed is in proportion to the intensity of the light which strikes a leaf, the smallest amount being in shady places; and the healthiness of a plant is *cœteris paribus* in proportion to the quantity of carbonic acid decomposed. Therefore, the healthiness of a plant should be in proportion to the quantity of light it receives by day. — [*Lind. The. of Hort.*]

suspended, as is the case in the artificial atmosphere of a hot-house. And as by artificial means the balance is therein destroyed, so, also, by artificial means must the elements of the atmosphere be adjusted, and the balance maintained.

It is impossible for us to contemplate so special an adjustment of opposite effects, without admiring this beautiful dispensation of Providence, extending over so vast a scale of being, and demonstrating the unity of the plan upon which the whole system of the organized creation is designed. And yet man, in his ignorance, has done his utmost to destroy this beautiful and harmonious plan. It was evidently the intention of the Creator, that animal and vegetable life should everywhere exist together, so that the baneful influence which the former is constantly exercising upon the air should be counteracted by the latter. Nothing, therefore, can be more prejudicial to the health of a large population, than the close packing of houses together, as presented in large cities. Hundreds of thousands of men, with manufactories of all kinds, — the smoke and vapors of which are still more injurious than the foul air produced by human respiration, — being crowded together in the smallest possible compass, with scarcely the intervention of an open space on which the light and air of heaven may freely play, and without any opportunity for the growth of any kind of vegetation sufficiently luxuriant to give pleasure to the eye, or sufficiently energetic to answer its natural purpose; for the close confined atmosphere of crowded cities is almost as injurious to vegetation as to animals; the smoke, which is constantly hovering above them prevents their enjoyment of the clear bright sunshine which they require for their health, and the dust, which is constantly floating in the atmosphere, covers the surface of their leaves, clogs up the pores, and prevents respiration.

This is the reason why plants thrive so badly in dwelling-houses in large cities, and also in the external air in the streets and squares. But lofty trees are so beneficial in such situations that they have with truth been called *the lungs* of large cities, so important is the effect produced by them in purifying the air. It is true, they may occasion some degree of dampness in the immediate neighborhood, but this evil is more than counterbalanced by the good they effect. "New Haven," justly called the City of Elms, is almost embowered in the shade of lofty trees, and is remarkable for the salubrity of its atmosphere, and the health of its inhabitants. There, almost every house has its garden; and the daily consumers of its deleterious exhalations stand in the open streets, at once the ornaments of the city and the scavengers of the air. The cutting down of a healthy tree, in the midst of a large town, without some very strong reason, should be regarded as an offence to the community, and an injury to the public weal. It is much to be wished that other towns, that are rapidly increasing in extent and population, would follow the example of New Haven, and bad ventilation and impure air would, in a very great degree, be deprived of their injurious effects.

2. Under favorable circumstances, plants are able to appropriate a larger amount of carbonic acid than that commonly existing in the atmosphere. The vegetation around the springs, in the valley of Gottingen, which abound in carbonic acid, is very rich and luxuriant, appearing several weeks earlier in spring, and continuing much later in autumn, than at other spots in the same district. But it is probable that, taking the average of the whole globe, and at all seasons, the quantity of carbonic acid existing in the air is that most adapted to maintain the health of the plants at present inhabitants on its surface, as well as to interfere as little as possible with the animal creation. In hot-houses, however, the case is different, especially in winter; for, although carbonic acid be not produced by the respiration of animals, it is produced in abundance by other causes, and these same causes also depriving the atmosphere of oxygen and its aqueous vapor, the carbonic remains in excess, and its effect upon the plants is easily perceived. The presence of oxygen, in proper quantity, in the atmosphere of a green-house, or hot-house of any kind, is even more necessary to be artificially maintained, than carbonic acid, because the oxygen affords the means by which the superfluous carbon is removed. We know that plants in a hot-house suffer more frequently from an excess of carbon than an excess of oxygen, arising from the causes

above stated. It has been calculated that hot-houses, during the application of fire heat, contain four times as much carbonic acid in their atmosphere as is necessary for the health of the plants.

"Charcoal possesses the property of absorbing some gases to a great extent, as may be seen by the following table, in which the numbers indicate the volumes of gases absorbed, that of the charcoal being taken as unity.*

Absorption of Gases by Charcoal.

Ammonia,	90	Bi-carb. hydrogen,	35
Muriatic acid,	85	Carbonic oxide,	9.4
Sulphureous acid,	65	Oxygen,	9.2
Sulphuretted hydrogen,	55	Nitrogen,	7.5
Nitrous oxide,	40	Carbur. hydrogen,	5
Carbonic acid,	35	Hydrogen,	1.7"

The above table will show how very useful charcoal may be rendered as an agent in the absorption of these gases, when present in excess, either in a plant-house or other places.

3. The evolution of heat by plants is most evident at those periods of their existence in which an extraordinary quantity of carbonic acid is formed and given off. This is the case during the germination of seeds; and though the heat produced by a single seed is too soon carried off by surrounding bodies to be perceptible, it accumulates to a high degree, where a number are brought together, as in the process of malting, when the thermometer has been seen to rise 110°. An extraordinary amount of carbonic acid has been found to accumulate in a hot-house, in one night, so as sensibly to affect the respiration of individuals entering the house in the morning; which shows the necessity of night ventilation. The disengagement of carbonic acid has been sensibly found in some plants, by the evolution of heat in some of their organs. Thus, the flower of a geranium has been found to possess a heat of 87°, when the air around it was 81°. As in the case of seeds, however, the production of heat is most sensible where the flowers are crowded together, and in those flowers where the size of the fleshy disk is

22* * Daniel's Introduction to Chemistry.

most considerable, the quantity of carbon to be united with the oxygen is consequently the greatest. And the combination of this cause with the other, causes the temperature of the clusters to be raised very high. A thermometer placed in the centre of five spadices has been seen to rise to 111°, and one in the centre of twelve, to 121°, while the temperature of the external one was only 66°.

From what has been stated, we think it may be argued that plant-houses require to be ventilated at night even more than during the day; but the quantity of air then admitted must be in proportion to the mean of the internal and external temperatures; but more particularly depending on the condition of the plants.

4. Various theories have been propounded by physiologists regarding the power of plants to withstand vicissitudes of temperature, and, among others, we have the following from the high authority of Decandolle : —

First, in the inverse ratio of the quantity of water they contain; secondly, in proportion to the viscidity of their fluids; thirdly, in the inverse ratio of the rapidity with which the fluids circulate; fourthly, in proportion to the size of the cells, so is the liability of the plants to freeze; fifthly, the power of plants to resist the extremes of temperature is in exact proportion to the amount of confined air which the structure of the plants enables them to contain. These and other principles are laid down, and, apart from their practical observation, they are of themselves sufficient to form the ground of theory. There is nothing, however, in the above calculated to be of material service to the gardener in the culture of exotic plants. The distinctions upon which rest their powers to resist changes of temperature are by far too undefined and minute to enable us to determine the quantity or quality of the organic elements they contain. Neither can we ascertain the dimensions of the cells with sufficient accuracy to determine the precise degree of heat or cold which any given plant will endure. In the management of tender plants, we must find a firmer foundation on which to rest our principles of action. We must endeavor to ground our judgment upon broader and safer principles; and, in order to reach this point, let us briefly consider the nature of atmospheric action upon hot-houses.

Before entering upon any illustration of its practical effects upon these structures, we will give an extract from Dalton's Chemical Philosophy, which will enable us to account more clearly for some of those results that we have often observed, and which have so often humiliated our practical pride and baffled all our boasted experience. In fact, they have been considered as belonging to that class of unaccountabilities which our Creator has placed beyond the ken of human discovery.

"It is a remarkable fact," Dalton observes, "and has never I believe been fully or satisfactorily accounted for, that the atmosphere, in all places and seasons, has been found to decrease in temperature as we ascend, and nearly in arithmetical progression. Sometimes this fact may have been otherwise, i. e., that the air was colder at the surface of the earth than above; particularly at the breaking up of a frost, I have observed it so. But this is evidently the effect of a great and extraordinary commotion in the atmosphere, and is generally of very short duration. What, then, is the occasion of this diminution of temperature in ascending? Before this question can be solved, it may be necessary to consider the defects of the common solution. Air, it is said, is not heated by the direct rays of the sun, which passes through it as a transparent medium, without producing any calorific effect till they reach the surface of the earth. The earth, being heated, communicates a portion to the atmosphere, while the upper strata, in proportion as they are more remote, receive less heat, forming a gradation of temperature similar to what takes place along a bar of iron, when one of its ends is heated." The first part of the above solution is probably correct. Air, it would seem, is singular in regard to heat; it neither receives nor discharges it, in a radiant state. If so, the propagation of heat through air must be opposed by its conducting power, the same as in water. Now, we know that heat, applied to the under surface of a column of water, is propagated upward with great velocity, by the actual ascent of the heated particles; it is equally certain that heated air ascends in the same way. From these observations, it would follow that the causes assigned above for the gradual changes of temperature in a perpendicular column of atmosphere, would apply to a state of temperature the very reverse of the fact; namely, that the higher the ascent, or the more distant from the earth, the higher would be the temperature. Whether this reasoning be correct, or not, we think it must be universally allowed that the fact has not hitherto received a very satisfactory explanation. We conceive it to be one involving a new principle of heat; by which we mean, a principle which no other phenomenon of nature presents us with, and which is not at present recognized as such. We shall endeavor, in what follows, to make out that principle.

The principle is this. The natural equilibrium of heat, in an atmosphere, is when each atom of air, in the same perpendicular column, is possessed of the same quantity of heat; and, consequently, the natural equilibrium of heat in an atmosphere is when the temperature gradually diminishes in ascending. That this is a just consequence cannot be denied, when we consider that air increases, in its capacity for heat, by rarefaction; and, therefore, if the quantity of air be limited, it must be regulated by the density. It is an established principle, that every body on the surface of the earth, unequally heated, is observed constantly to tend towards an equality. The new principle announced above would seem to suggest an exception to this law; but if it be thoroughly examined, it can scarcely appear in that light. Equality of heat and equality of temperature, when applied to the same body, in the same state, are found to be so uniformly associated together, that we scarcely think of making any distinction between the two expressions. No one would object to the commonly observed law being expressed in these terms. When any body is equally heated, the equilibrium is found to be restored, when each particle of the body becomes possessed of the same quantity of heat. Now the law, thus expressed, is what I apprehend to be the true general law, which applies to the atmosphere as well as to other bodies. It is an equality of heat, and not an equality of temperature, that nature tends to restore.

The atmosphere, indeed, presents to us a strikingly peculiar feature, in its regard to heat. We see, in a perpendicular column of air, a body without any change of form, slowly and gradually changing its capacity for heat, from a less to a greater; but all other bodies retain a uniform capacity throughout their substance. If it be asked why an equilibrium of heat should turn upon the quality in *quantity*, rather than in *temperature*, I answer, I do not know; but I rest the proof of it upon the fact of the inequality of temperature observed in the atmosphere in ascending, which invariably becomes colder as we ascend in height; while, in artificial atmospheres, as in the case of a hothouse, the fact is quite the reverse. If the natural tendency of air was to an equality of temperature, there does not appear to me any reason why the lower regions of air are warmer than the higher, or why the law of equalization held good in one case and not in another.

To enable us to apply these arguments more clearly to our subject, it will be necessary more fully to consider the relation of the atmosphere in regard to heat; and the arguments already advanced in behalf of the principle we are endeavoring to establish, are powerfully corroborated by the following facts.

We find, by the observations of Bougeur, Sassure, and Gay Lussac, that the temperature of the atmosphere, at an elevation where the weight is half that at the surface, (about 14,000 feet, or less than three miles,) is reduced in temperature 50° Fahrenheit; and, from experiment, it appears that air, suddenly rarefied from two to one, produces 50° of cold. Hence we might infer that the stratum of air at the earth's surface being taken up to the height above mentioned, preserving its original temperature and suffered to expand, becomes two measures, and is reduced to the temperature of the surrounding air, and *vice versa*. In like manner, we may infer, if a column of air from the higher strata of the atmosphere were condensed and brought into a horizontal position on the earth's surface, it would become of the same density and temperature as the air around it, without receiving or parting with any heat whatever. Another important argument in favor of the theory here advanced, may be derived from the contemplation of an atmosphere of vapor. Suppose the present aerial atmosphere were to be substituted for one of aqueous vapor; and suppose, further, that the temperature of the earth's surface were uniformly 212°, and its weight equal to 30 inches of mercury. Now, at the elevation of about six miles, the weight would be fifteen inches, or one half of that below; at twelve miles, it would be $7\frac{1}{2}$ inches, or one fourth of that at the surface, and the temperature would probably diminish 25° at each of these intervals. It could not diminish more; for the diminution of temperature 25° reduces the force of vapor one half. If, therefore, a greater reduction of temperature were to take place, the weight of the incumbent atmosphere would increase, being converted into water, and the general equilibrium would thus be disturbed by condensation in the upper regions."

It has been observed, that if the ventilators of hot-houses be kept close during the day, the internal temperature will rise, although no artificial heat be applied; from which it has been supposed that glass freely admits the calorific rays to pass through it, in their descent, but arrests it in their upward progress. Professor Robinson has proved that glass freely transmits the luminous rays, but stops the calorific rays, till it becomes saturated with heat to a certain degree; which proves, also, that light and heat are not identical, although both obey the same laws of reflection, refraction, and radiation. Although heat may arise from the same source as light, and possess a great affinity to it, yet caloric possesses properties peculiar to itself, and differs in its degree of affinity for other bodies; for, although it has a tendency to come to an equilibrium, when bodies differing in quality are exposed to its influence, it has been found that these bodies do not all come into an equal temperature at the same time. Caloric readily enters into some bodies, and freely combines with them, whereby their temperature becomes increased, and their properties sometimes changed. [See Part I., *Construction*, sec. Glass, p. 106.]

From the foregoing remarks, it will easily be perceived that many of our operations, in the management of hot-houses, are not only theoretically wrong, but diametrically opposed to the laws of nature. Our methods of ventilation are wrong in practice, because our notions are wrong in principle. We raise the temperature of our houses by artificial means, and drive off the oxygen and aqueous vapor, without returning a supply. We admit the heat to escape through the laps and fissures of the glass, of which there is always enough in badly glassed houses to admit the escape of one fourth the heat radiated in the house. And, moreover, we allow one fourth more at least to be taken away by direct radiation from the glass, so that hardly one half of the heat generated is used for the purpose intended. And, lastly, we admit the external air into the house, to deprive the atmosphere of its moisture by condensation. Likewise, in summer, we admit the external air, in Sirocco currents, to sweep through the house, carrying away the moisture daily by gallons; and which, if not returned in equal abundance, must speedily prove injurious to the plants.

SECTION II.

1. The ventilation of hot-houses, during winter, requires all the skill which the most experienced gardener has at command. It is a comparatively easy matter to open and shut the sashes, or ventilators; but to do so with benefit to the plants, at all times, requires an amount of skill which is seldom bestowed upon it. Admitting large quantities of cold air into a house, many degrees below the internal atmosphere, cannot be otherwise than injurious to the plants growing therein. It has been calculated that a volume of air, equal to 400 cubic feet, will absorb upwards of 36 gallons of water during its rise from 60 to 90° of temperature; or, in other words, upwards of a gallon has been absorbed for every degree of temperature above 60°. This will, in some measure, show the propriety of keeping the walls and floors of a plant-house continually saturated with moisture, especially during the hot days of summer, as well as of preventing currents of air from sweeping through the house. We have succeeded, in this way, in keeping the atmosphere of a green-house 10 or 15° below the external temperature, even when the latter stood above 90°; and almost every gardener, who has paid attention to these matters, has experienced the same results. It is a common error for gardeners to give large supplies of air, in sultry weather; but, as it is a practical one, and one of long standing, it is excusable in those who have not studied its effects attentively.

By far the larger number of gardeners attach great importance to the ventilation of their houses abundantly, without perhaps sufficiently considering the nature of the plants they have to manage; and, as has been justly enough said, by supposing that plants require to be treated like man himself. They consult their own feelings, rather than the principles of vegetable growth. There can be no doubt, however, that the effect of excessive ventilation is more frequently injurious than advantageous; and that many houses, and especially hot-houses, would be more skilfully managed, if the power of ventilation possessed by the gardener were much diminished.

Animals require a continual renovation of the air that surrounds them, because they very speedily render it impure, by the carbonic acid given off, and the oxygen abstracted by animal respiration. But the reverse is what happens to plants. They exhale oxygen during the day, and inhale the carbonic acid of the atmosphere, thus depriving the latter of that which would render it unfit for the sustenance of the higher orders of the animal kingdom ; and, considering the manner in which glass-houses of all kinds are constructed, the buoyancy of the air, in all heated houses, would enable it to escape in sufficient quantity to renew itself as quickly as it can be necessary for the maintenance of the healthy action of the organs of vegetable respiration.

It is, therefore, improbable that the ventilation of houses, in which plants grow, is necessary to them, so far as respiration is concerned. Indeed, Mr. Ward has proved that many plants will grow better in confined air, than in that which is often changed. By placing various kinds of plants in cases, — not, indeed, air-tight, for that is impossible with the means applied to the construction of a glass-house, but so as to exclude as much as possible the admission of the external air, — supplying them with a due quantity of water, and exposing them fully to the light, he has shown the possibility of cultivating them without ventilation, with much more success than usually attends glass-house management.

2. In forcing-houses, in particular, it will be evident from what is about to follow, that ventilation, under ordinary circumstances, in the early spring, may be productive of injury rather than of benefit. Many gardeners now admit air very sparingly to their vineries during the time that their leaves are tender and the fruit unformed. Some excellent hot-houses have no

23

provision at all for ventilation; and we have the direct testimony of Mr. Knight, as to the advantage of the practice to many cases to which it has been commonly applied.

"It may be objected," says Knight, "that plants do not thrive, and that the skins of grapes are thick, and that other fruits are without flavor, in crowded forcing-houses. But in these, it is probably light, rather than a more rapid change of air, that is wanting; for in a forcing-house, which I have long devoted to experiments, I employ but very little fire heat, and never give air till the grapes are fully ripe, in the hottest and brightest weather, further than is just necessary to prevent the leaves from being destroyed by excess of heat. Yet this mode of treatment does not at all lessen the flavor of the fruit, nor render the skins of the grapes thick. On the contrary, their skins are always moist, remarkably thin, and very similar to those grapes which have ripened in the open air." — [*Hort. Trans.*]

We have experienced the same results, as those recorded by Mr. Knight, under similar treatment, and that too under a more powerful sun. We have pursued this method of giving a very limited supply of air, on an extensive scale, in some large graperies in Maryland, and under glass of the very worst possible description. Yet, during one of the hottest summers which had been experienced for some years, these vines grew beyond anything we had ever seen, without any indication of injury by the sun's burning rays. The lower surfaces of the houses, however, were kept moist, by frequent sprinkling with water during the day. Many large houses in England are never aired, except, perhaps, a few apertures at the top of the house, which are left open, night and day, during the summer. But in all cases within our knowledge, water is abundantly supplied to the atmosphere from the floors of the house.

3. The philosophy of this method is easily perceived. The under surface of the glass is continually covered with a deposition of the evaporated moisture, which intercepts the calorific rays, and prevents them from being concentrated on the leaves, from which cause the leaves are scorched and burned; the atmosphere, at the same time, undergoing comparatively little change, or admixture with the external air.

While, however, the *natural* atmosphere of a hot-house cannot be supposed to require changing, in order to adapt it to the respiration of plants, it is to be borne in mind that the air of hot-houses, artificially heated, may be rendered impure by the means employed to produce heat, as will be seen from what has already been said on the principles of heating, in the preceding part of this work. Sulphuric acid gas, in variable quantities, escapes from brick flues, especially old and imperfectly constructed ones, and various other unsuspected sources of impurity, an infinitely small quantity of which is sufficient to contaminate the air, in respect to vegetable life.

Drs. Turner and Christison found that $\frac{1}{10000}$ of sulphurous acid gas destroyed leaves in forty-eight hours; and similar effects were observed from hydro-chloric acid gas. Chlorine, ammonia, and other gases produce the same results, when their presence is altogether undiscoverable by the olfactory organs. We also know that the destructive properties of air, poisoned by corrosive sublimate, by its being dissolved and evaporated in the atmosphere of a hot-house, is not appreciable to the senses. [See Chemical Combinations in the Atmosphere, sec. IV., for detailed information on this subject.]

Ventilation is necessary, then, not to enable plants to exercise their respiratory functions, provided the atmospheric air is unmixed with accidental impurities, but to carry off noxious vapors generated in the atmosphere of a glazed house, and to produce dryness, or cold, or both. Thus it is evident that air is given under many conditions, when it is not only unnecessary, but injurious.

When air is admitted, to produce cold in the house, the external temperature must be lower than the atmosphere of the house. This effect, however, cannot always be produced by ventilation, as, in summer, if the houses be rightly managed, the reverse effect will be produced, as the external air is not only warmer but more drying in its nature than the air of the house ought to be; therefore its admission can only prove injurious, rather than otherwise, as we shall afterwards show. On the other hand, if the external air be cold, its admission will produce dryness, which may also prove injurious under certain circumstances.

4. When the external air is admitted into a glazed house, below the temperature of the air it contains, the heated moist air rushes out at the upper ventilators; or, if it cannot find egress, it is quickly condensed upon the cold surface, against which it is forced to ascend; the latter rapidly abstracts from the plants, etc., a part of their moisture, and thus gives a shock to their constitution which cannot fail to be injurious.

This abstraction of moisture is in proportion to the rapidity of motion in the air. But it is not merely dryness that is thus produced, or such a lowering of the temperature as the thermometer suspended in the house may indicate. The rapid evaporation that takes place, upon the admission of the air, produces a degree of cold upon the surface of the leaves, and of the pots in which they grow, as well as all other bodies around them, of which our instruments give no indication. To counteract these mischievous effects, many contrivances have been proposed, in order to insure the introduction of fresh air, warm and loaded with moisture; such as compelling the fresh air to enter a house, after passing through pipes moderately heated, or over hot-water pipes surrounded by a damp atmosphere, which have been proved decidedly advantageous, and to which we will subsequently refer.

If ventilation is merely employed for the purpose of purifying the air, i. e., for carrying off extraneous gases and vapors that may be generated by artificial heat, it should be introduced, by all means, with great caution; and some expedient should be adopted for supplying it with moisture, as well as to warm it slightly on its passage inwards, more especially in cold, frosty, or windy weather.

If it is only introduced for the purpose of lowering the temperature, as in mild and genial spring and summer weather, it may be admitted without any such precaution; and the freedom of admission should be in proportion as the external and internal temperatures approach each other in equality.

In hot, sultry weather, air should be sparingly admitted, as the same effects are produced by the excessive evaporation as by the currents of cold air, as will be afterwards shown.

5. Ventilation is also required, in winter, in pits and frames where soft and succulent plants are grown, especially in pits and frames warmed with fermenting materials. In this case, much care and caution are necessary; the object here being to carry off the superfluous moisture, in order that the succulent tissue of the plants may not absorb more aqueous matter than they can decompose and assimilate. Although these kinds of plants will bear a high degree of atmospherical moisture in summer, when the days are long and the sun bright, and when, consequently, all their digestive energies are in full activity, yet they are by no means able to endure the same amount in the dark, short days of winter, when their powers of decomposition, or digestion, are comparatively feeble.

6. The thermometric changes are by no means satisfactory guides for regulating the admission of air in hot-houses, as the effect required by the indications of the thermometer may be produced without resorting to the admission of air. In hot-houses, we have full control over the state of the atmosphere, both as regards its moisture and temperature; and the means of exercising this power ought to be known and familiar to every gardener. But there are many circumstances which ought to be duly considered in the exercise of this power, and some unsuspected results arise from the unlimited use and exercise of it; and, as has been already said, by far the greater number of gardeners attach too much importance to the mere opening and shutting of sashes, windows, etc., without duly studying the rationale of the practice. We will show that the practical effects of ventilation are not only different from what many suppose, but are actually injurious.

During winter we are in the habit of raising the temperature of our hot-houses, by artificial heat, to 45 or 50°; then, for six or seven hours during the day, we open the lights and admit a large quantity of cold air. This is also a stumbling-block, on which a great many gardeners fall; for it is not solely to the

23*

temperature, but rather to the hygrometrical state of the atmosphere, we ought to look. We ought to regulate the admission
of air, not solely by the thermometer, but also by the hygrometer; for, upon the latter condition, the health of the plants, and
the perfection of their flowers and fruit, very much depend; —
and, consequently, it is a matter which ought to be studiously
considered. Nothing is more injurious than the admission of
currents of air when the external temperature is lower than the
internal one; and more especially so to plants that have been
for a considerable time subjected to a high temperature by artificial heat.

The causes which operate in rendering the atmosphere of
hot-houses unnaturally arid may be said to be two-fold. The
first is the condensation of moisture upon the glass, arising
from the action of the external cold upon its upper surface. The
second is the escape of heated air through the laps and crevices of the glass, and otherwise. This heated air escaping, carries along with it a large quantity of contained moisture, the
loss of air being supplied with cold, dry air, which finds access
by the same means. The loss of heat and moisture sustained
by these means is far more than would be supposed by those
who have not calculated the amount.

7. We have seen that the quantity of moisture a cubic foot
of air will hold in invisible suspension depends on its temperature; and as the temperature is increased, so is its capacity
for moisture. Suppose, then, that this capacity is doubled
between the temperature of 40 and 60°; that is to say, every
cubic foot of air that enters the house at 40°, and escapes at 60°,
carries with it just double the quantity of moisture it brought in.
Now, every one must be sensible that these circumstances, continued for any length of time, must render the atmosphere of
the house too arid for healthy vegetation; and, consequently, if
the deficiency of moisture so occasioned be not supplied by
artificial evaporation, then the plants must part with their secretions to supply the atmospheric demand, and the soil and other
materials in the house will also be drained of their moisture, to
make up the deficiency. The greater the difference between

the internal and external temperature, the greater will be the demand for moisture. Thus, if the external air be at the freezing point, (32°,) and the air in the house heated to 50 degrees, then there is three times more moisture carried away by escaping air than is brought in by the returning quantity; and, escaping at 90°, it carries away four times as much, and so on, in proportion to the difference of the two atmospheres; the external air, however, increasing in ratio as it decreases in temperature.

According to these calculations, atmospheric air, entering a house at 32°, and escaping at 100°, carries away nearly six times as much moisture as it brings in. This, in a short time, would render the atmosphere of a house deleterious to either animal or vegetable life; and in large and lofty houses this is practically the case. We have managed a lofty plant-house, where the plants on the side shelves were nearly frozen, while the thermometer, hung in the angle of the roof, about 45 feet high, stood at 100 degrees. Now this heated air, escaping at the top of the roof, as is generally the case as well as here, carried away more moisture than the small evaporating surface could supply; the effects were, consequently, ruinous to the plants. However imperfect the above calculations may be, they are within the bounds of truth, and are sufficiently accurate to show the importance of this subject to exotic horticulture; and it will more effectually impress upon our minds the amount of care and consideration which the ventilating of hot-houses demands. If air *must* be admitted, for the purpose of regulating the internal temperature, every precaution should be taken to prevent it from entering in strong currents, and it should be taken in from the warmest side of the house, and, if possible, over a warm surface, — as hot-water pipes, or whatever heating apparatus may be employed, — so that the internal atmosphere may be gradually reduced; and, at the same time, the utmost precaution should be used to prevent the escape of heated air, at least as little as possible, by direct ascension; this is easily accomplished by the improved methods of ventilation now adopted, some of which I shall hereafter endeavor to describe. Thus the cultivator is enabled to modify the two atmospheres, previous to their com bination, and by raising the humidity in the atmosphere of the house, to compensate for that carried away by the egress of heated air, the plants will breathe an atmosphere more conducive to their healthy development, and will be benefited by the change.

S. Every gardener has observed the water on the under surface of the glass, in the morning, before the sun has risen, warmed the glass, and driven it off again, in the form of aqueous vapor. This affords us a good illustration of the immense quantity of moisture carried upwards by the heated air, and deposited upon the glass, by condensation. This moisture is, of course, taken away from the plants, and other bodies capable of giving it off, and is demanded by the air as it becomes warm, and capable of carrying a larger quantity than when no fire was applied, — or rather, when the temperature of the house and the temperature of the external air were alike, for in such case no condensation on the glass would take place ; and, as I have remarked, the proportion of water deposited will be in exact ratio to the intensity of the external cold ; thus, the greater the difference, the greater the deposition ; for then the action of the external cold upon the upper surface of the glass being greater, and the two atmospheres being brought into more rapid proximity, the particles of heated air are cooled as quickly as they ascend to the under surface of the glass; they then fall to supply the place of others, leaving the contained moisture upon the cooling surface, in the form of dew, — the same process being repeated through the whole night, or until an equality of temperature is established ; the quantity thus deposited amounts to immense volumes of water.

9. Experiments have proved that each square foot of glass contained in the roof of a hot-house will cool down $1\frac{1}{4}$ cubic feet of heated air per minute as many degrees as the temperature of the internal exceeds that of the external atmosphere. Suppose, for instance, that the external air stands at 40°, and that of the house 60° ; then, for every square foot of glass contained in the house, one and one fourth cubic feet of

air will be cooled down the 20 degrees ; thus, 60 minus 40 gives the difference, which is 20. If the house contains 800 square feet of glass, presented to the action of the external atmosphere, 1000 cubic feet of air will lose 20 degrees of heat ; consequently, the moisture this air held in invisible solution, in virtue of its 20 degrees of temperature, will be condensed by the external cold, and deposited on the glass ; and it will also be found, that the greater the difference between the external and internal temperatures, the greater will be the amount of condensation. The quantity of moisture abstracted from plants, at high temperatures, is enormous. This fact is sufficiently demonstrated in a hot summer day, when the leaves of the trees are wilted, and the garden vegetables flag and droop their leaves. The earth gives out its moisture, and the atmosphere carries it away. The same thing takes place in hot-houses ; the moisture is abstracted by the heated air, and is carried off in the form of invisible vapor, till its upward progress is arrested by the glass, and the cold again reduces it to water.

If we take, for example, the roof of a hot-house, comprising 750 superficial feet of glass, and calculate that every square foot of that glass will cool down $1\frac{1}{2}$ cubic feet of heated air 36° per minute, and calculating the internal temperature at 65°, we shall find that 937 cubic feet of air will be cooled down 36 degrees per minute. Now air, saturated at the temperature of 65 degrees, contains about 6·59 grains of water per cubic foot, and at the temperature of 30 degrees, it is saturated with 2·25 grains ; this gives 4·34 grains of water lost, in condensation on the glass, per minute ; or further, each square foot of glass condenses $1\frac{1}{4}$ cubic feet, or about 5·42 grains of water, per minute ; and supposing the atmosphere of a house, such as we have described, to be constantly supplied with moisture, by evaporation, or otherwise, there would be abstracted from it about $\frac{7}{8}$ of a pint of water per minute, which is about 12 quarts per hour, or at the rate of nearly 72 gallons in 24 hours. This enormous amount of water, evaporated into the atmosphere of a hot-house, when reduced to calculation, and displayed in plain figures, seems to startle the imagination, and looks very like exaggeration ; although it is much below the mark which, by a more

accurate calculation, it would certainly reach, yet the accuracy
of these calculations will appear sufficiently obvious to any one
who has paid studious attention to the subject.   I say studious
attention, because a person may be tolerably observant of atmospheric phenomena, and yet not form anything like an accurate
idea of this extraordinary process going on in his presence, and
the effect thereby produced on the vegetable system.   When
we enter a hot-house, on a cold, frosty morning, after a strong
fire has been kept up during the night, we are very apt to regard
the moisture condensed upon the lower surface of the glass
as an evidence of a healthy atmosphere and luxuriant vegetation ; and often have I heard it stoutly asserted, that it
was merely the effect of an excess of moisture in the atmosphere of the house.   This may be partly true, but the conclusions
which are drawn from the fact are founded on misconception,
that the moisture thus deposited on the glass has already **performed** its purpose of benefit to the plants.

SECTION III.

1. If we admit the truth of the foregoing calculations, (and we cannot justly reject them, until they are disproved by calculations more accurate, and observations more extended,) then we must acknowledge, also, that the old methods of ventilating hot-houses, which are still in common practice, are contrary to what we know to be right. Hence the question arises, How are these methods to be improved? Now, I would remark, that the mere *system* on which a house may be ventilated is of comparatively little importance, for no method of ventilation will be good, if the atmosphere be unskilfully managed. Various plans have been employed to modify the influence of draughts, or currents of air, many of which can hardly be termed improvements, since the general effect is the same as by the old method of opening the top and bottom sashes, which admits a current to rise up beneath the under surface of the glass, and, as it proceeds towards the aperture made by letting down the upper sashes, it carries the ascending moisture along with it, without in the slightest degree mixing with, or purifying, the volume of atmosphere contained in the lower portions of the house.

2. It has long been an object among gardeners to obtain a motion in the atmosphere of a hot-house; and to secure this, even machinery has, in some instances, been employed, and, under certain conditions of the atmosphere, these machines may go on very well. But subject to those vicissitudes of climate, so prevalent in many parts of this continent, the consequent result of their adoption is, a complete derangement of all that equalizing regularity which they were intended to secure. It appears to us a matter of considerable difficulty to lay down a definite rule, or propose a particular system of ventilating a house, since almost every locality has some characteristics pecu-

liar to itself. It is true, the elements of the atmosphere may be
nearly the same in one place as in another; but they are influenced by various circumstances, in different localities, and hold
soluble matters in suspension in very different proportions; and
in places much screened by trees, buildings, and similar objects
of shelter and obstruction, air may be admitted with greater
impunity than in situations exposed to wind from every quarter
of the compass, — the latter condition, as a matter of course, requiring more care, not only in the adjustment of the apertures
of admission, but also in the admission itself. The course of
the current of air, by the common methods of ventilation, — that
is, by opening the front, and letting down the top sashes, — is exceedingly variable ; sometimes the actual motion created in the
atmosphere is little more than a foot, or fourteen inches, below
the surface of the glass. This motion can be easily determined
by holding the flame of a candle in the current, when the flame
will incline towards the aperture of egress ; lower it gradually
down, till it assumes and maintains a perpendicular position,
being no longer affected by the current, the volume of air being,
in fact, stationary, except there be some aperture of ingress else-
where. We have found this simple operation exceedingly useful
in determining the currents of air in large houses, and, in most
cases, it seldom fails in giving an accurate indication of their
course.

However desirable a motion may be in the atmosphere of a
hot-house, — and I do not doubt but it is beneficial, — yet it is not
necessary that we should run headlong either upon Scylla or
Charybdis. There is a great difference between a motion in
the atmosphere created by the warm particles ascending, and
being replaced by the denser and colder air, and that created by
a tornado sweeping through the house. The former motion is
only perceptible to the eye of the attentive and experienced cultivator, and he can tell at a glance, by the quivering of the
leaves, that they are fanned by a gentle zephyr. I am aware
that some gardeners have a peculiar fancy for seeing their plants
and vine-leaves bristling about by a good wind, and may be
very successful, too, in their productions; but it cannot be asserted that it is compatible with a high state of gardening skill

or with that perfection in horticulture at which it is our duty to aim; inasmuch as the revelations of science are against it, as has already been shown, and practice has hitherto given no evidence to prove it beneficial to tender plants.

3. In large and lofty structures, and especially in dome-shaped houses, the management of the atmosphere becomes a matter of much more importance than in small houses. During mild and temperate weather, things may go on very well, as at such times the external air may be allowed to circulate through the house with greater impunity; but during the heat of summer, and the cold of winter, the atmosphere is much more difficult to equalize. With a frosty air externally, and the temperature at the surface of the earth down to zero, it is impossible to maintain a proper degree of temperature, in all parts of the house, without positive injury to those plants that may be growing, or have their branches extended into the upper regions of the house. In fact, without the precaution of covering, or some such expedient, mischief is absolutely unavoidable. What has already been said, upon the nature and properties of air, will sufficiently explain the cause; and, although it has been repeatedly asserted by theorists, that one part of a house being heated by radiation, from a body radiating heat, the equalizing law of nature will heat all parts of the house to the same temperature, and as speedily, too, yet we must enter our decided protest against the practical correctness of such a statement; at least, in our own practice, we have never found it so, under any circumstance, or by any system of heating. And hence, whatever the natural law of equality may be, the practical effects cannot be mistaken, or disputed, as far as regards hot-houses. We know that heated bodies tend to an equality of temperature; but, as has been already observed, air, of all other bodies, possesses peculiar properties in this respect in regard to heat, and in nothing is this peculiarity more strikingly illustrated than in the case under consideration.

4. With regard to the motion of the atmosphere in a hot-house, we know that the greater the difference between the tem-

24

perature of the air entering the house and the atmosphere of
the house itself, the greater will be the movement produced
among the particles. The motion is in exact proportion to the
difference of temperature ; and hence the necessity of admitting
the external air, in small quantities, when the external thermometer is low. The slightest cause that disturbs the equilibrium of the air produces a motion. It is more sensible than
the most delicate balance. It is put in motion by the slightest
inequalities of pressure, and by the smallest change of temperature. It is speedily rarefied by heat, and thereby rendered
specifically lighter than the neighboring portions, so that it
descends, while colder, and consequently denser, flows in, to restore the equilibrium. It will be easily seen, from the very
nature of this law, that an equilibrium cannot be maintained in
the artificial atmosphere of a hot-house, since the source of
radiation must necessarily be confined to too small a surface to
equalize the ascending heat ; and, on the other hand, the condensation by cold is too irregular throughout the heated volume. This irregularity, produced by its unequal action on
different parts of the house, must ever render it impossible to
obtain an equality of temperature throughout an atmosphere
heated by artificial means ; and the larger the house, the greater
will be the difficulty of maintaining an equilibrium in its various
parts. So much so is this the case, that, as has been already
stated, the difference has been found to amount to 100°.

"Gaseous bodies expand equally for an equal increase of
temperature, as measured by the thermometer. Gay Lussac
showed that 100 measures of atmospheric air, heated from the
freezing to the boiling point, became 137.5 measures ; consequently, the increase for 180° Fahrenheit is $\frac{37.5}{100}$ of its bulk.
Dividing this quantity by 180, we find that a given quantity of
dry air expands $\frac{1}{480}$ of the volume it occupied at 32°, for every
degree of Fahrenheit. New experiments have been made by
Rudberg, within a few years, giving $\frac{1}{491}$ as the ratio of expansion for one degree of Fahrenheit ; and these results are confirmed by Regnault. This last number may be adopted as the
true increment.

"If we wish to ascertain the volume which 100 cubic inches

of a gas at 40° would occupy at 80°, we must remember that it does not expand $\frac{1}{491}$ of its bulk at 40° for each degree, but $\frac{1}{491}$ of its bulk at 32°. Now, 491 parts of air at 32° become 492 at 33°, become 493 at 34°, and so on. Hence we can institute a proportion between the volume at 40° and that at 80°. *

Vol. at 40°.	Volume at 80°.		Cubic inch.		Cubic inch.
491 + 8	491 + 48	: :	100	:	108

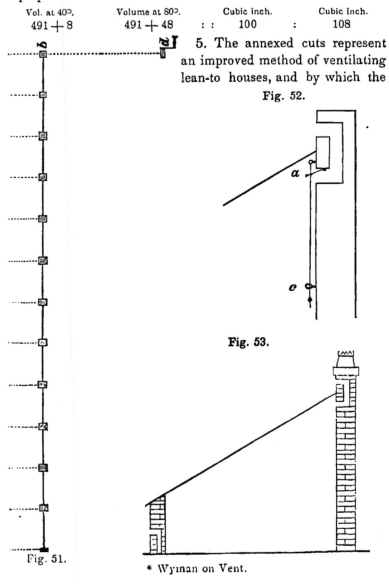

5. The annexed cuts represent an improved method of ventilating lean-to houses, and by which the

Fig. 52.

Fig. 53.

Fig. 51.

* Wyman on Vent.

whole house may be aired in the space of one minute; or as many houses as may be in the range. This is effected by a rod passing along the whole length of the house. A pulley is fixed immediately above each ventilator, and another placed opposite it upon the rod, as shown in Fig. 51. A piece of chain or cord is attached to the ventilator at one end; and passing over the pulley, as shown at *a*, Fig. 52, is then fixed to the pulley placed opposite it upon the rod. A larger wheel, or pulley, is fixed at one end of the rod, (*b*,) to which is attached a chain, connected with a crank, situated within the reach of a person standing on the floor. This crank is fixed on the back wall, as seen at *c*, Fig. 52.

From the foregoing cuts and description it will be perceived that, by giving the crank (*d*) a few turns, the whole of the ventilators will be opened. The crank is provided with a racket, so that they may be opened to any distance, from half an inch to the full height.

The ventilators in the front wall may be opened and shut by the same method, and may be, for convenience, brought from the outside. Any length of house, or any number of houses, may be ventilated at once by this method, providing the apertures are in a straight line; their perpendicular distance from the horizontal shaft makes no difference in their facility of working. The pulley cords of the higher ones only require to be lengthened according to the distance, the diameter of the wheel on which the cord turns being equal all along the shaft.

6. Figures 54 and 55 represent a method of ventilating span-roofed houses. It is employed in the houses at Frogmore,

Fig. 54.

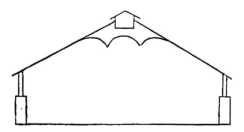

in England. Fig. 54 represents the end section of the house, with the ventilator in proportion to the other parts. Fig. 55

Fig. 55.

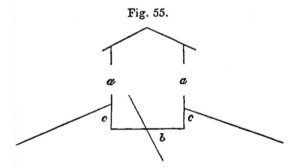

shows the sectional view of the ventilator, enlarged: *a a* are openings of admission, and are covered with lattice-work, to break the force of the current of ingress; *b*, the movable shutter, which regulates the admission to and egress from the house. It is scarcely necessary to observe that these houses have been ventilated on the most approved principles; and it appears that several advantages are gained by this method. For instance, the current of heated air is arrested, in its progress outwards, by the depending glass at *c c*, and is, in some measure, thrown downwards, preventing also the escape of its contained moisture. There is no doubt this method is very commendable for span-roofed houses; and one of its advantages is, that the house can be aired, at any time, without the plants being saturated with rain.

It is very possible that these compound systems of ventilation may excite a smile from some who have, all their lifetime, been accustomed to pull heavy sashes up and down for the purpose of giving air. But if we include, in one computation, the labor, the time, and the advantages of giving a range of houses three or four hundred feet long, air at the proper time, and all at the same moment, we will find a value in the system worthy of something more than the mere smile of passive silence, which is too frequently all that is at first accorded to such improvements.

In some establishments, instead of pulleys, toothed wheels are fixed to the shaft, which are made to work in a curved handl-

24*

attached to the front sash by means of a hinge. This curved rod is toothed on the lower side to answer the wheel, and is kept in its place by an iron staple, having an eye through which the sash-handle passes, as seen at *a*, Fig. 56. A crank and rachet-

Fig. 56.

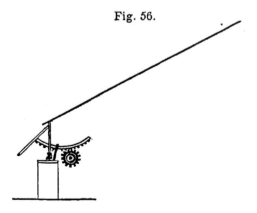

wheel is provided, at one end of the shaft, by turning which the sashes are simultaneously opened and shut, to any distance. This method is simple and efficient. It has been extensively carried out in the unique assemblage of horticultural buildings at Frogmore; and, as an improvement in the modes of ventilating hot-houses, is considered, by competent judges, the most valuable contrivance that has been introduced during the last half century. By the turning of a small windlass, (which any child may do,) any quantity of air may be admitted, and increased or diminished at pleasure, throughout the whole range of buildings.

The ventilation of forcing-houses, by this compound method of opening the whole sashes at once, is very liable to produce serious results, before the person in charge becomes fully acquainted with the management of it. This, like many other really valuable improvements in gardening, has been adopted,— bungled in the construction, — mismanaged afterwards, — then, lo! it is condemned, with all the pomp and dignity of *practical experience!* The present moment affords an ocular demonstration of this too common fact. Some people suppose, if they can only get mechanical contrivances to accomplish certain ends that all is right. It is certainly desirable to employ mechanical contrivances, whenever they can, as in the present case, be applied advantageously. But mechanism can never make a gardener, inasmuch as the chief part of what constitutes a real gardener springs from mental, not physical, activity. It is a very easy matter to open and shut the ventilators of a hothouse; but it requires something more than mere mechanical power to do so with certain benefit to the inhabitants within.

This will be rendered clear by a common illustration. Let a dwelling-room be warmed to a temperature of 60°; and suppose it to be tolerably well filled with individuals, by the animal heat and respiration of whom the room by and by becomes somewhat raised in temperature, and contaminated in its atmosphere. Then, all at once, let the windows be thrown open, and the consequence is not only disagreeable, but highly dangerous, as is manifest by the murmur which very soon pervades the assembled party. Now, the case is precisely similar in a hot-house, only with this difference, — the unfortunate plants cannot speak in audible sounds to tell the injuries that are perpetrated upon them; yet they bear a language, imprinted on their leaves, no less truthful, nor less understood by the attentive observer. The above common occurrence is a plain illustration of what I have often seen, and have been forced to perform, in the ventilation of forcing-houses, and which is more likely to be exemplified by the compound methods which I have described. Science may enable us to be more watchful of atmospheric phenomena, and may draw our attention to facts which mere practice might pass unnoticed. But this is a practical operation which science has not yet approached, and which, in all her discoveries, she never can approach, *i. e.*, to tell us the precise quantum of air to admit at different times and under different temperatures. The method of mixtures does not come near it, and the combination of gases gives the gardener little scientific assistance. We *must* know the nature and properties of air at all times and temperatures; but the quantities and proportions in which we are to admit it must be learned by experience and strict observation. We must watch its effects upon the plants, and admit it **in** proportions which appear, by oft-repeated trials, to be **most** beneficial.

7. We could describe several other systems of ventilation, by what we have called the compound method, which have a greater number of wheels and rachets, and other kinds of machinery about them, but which possess no advantage over either of the methods we have described. One system, in particular, has received some countenance, which consists in opening by the aid of a spring instead of the toothed rod, as shown in Fig. 56. We have managed various houses ventilated by this method, but we must say that it worked badly, although much care had been taken to have the machinery properly fitted up; for instance, where the springs are of unequal strength, — and by constant use they very soon become so, — you will find a very great irregularity in the airing of the house, some of them requiring to be opened nearly full length, before the others will open a few inches. Again, if some of the sashes be stiff to open, those that are not so will open freely, while the ones that are hard to move will not open at all. This has frequently caused us much annoyance. It can never occur with the toothed wheel, as an equal force is exerted on each ventilator or sash, and every sash is opened to a regular distance. But if any of the sashes be stiff to open, then the whole power applied is directed upon them alone, until the whole move together. The only supposed advantage of the springs is, that they do the work silently, whereas a little noise is made by the rachet-wheels, — a matter, in most cases, of so trifling importance, as to be unworthy of consideration; but, as drowning men catch at straws, so the most insignificant circumstance is eagerly seized, and magnified into momentous import, by would-be inventors, for the purpose of palming off their so-called invention upon the community, and sustaining its sinking reputation. The less machinery there is about a hot-house, the better; and that system which does its work in the most efficient manner, with the smallest amount of labor, and is least likely to get out of order, is decidedly to be preferred. This is a commendation which cannot be justly given to some late inventions; and, without wishing to throw

anything in the way of improving our present systems, or discouraging the application of new mechanical inventions to aid the practical operations of horticulture, we would say that some of these methods lately brought into notice may be justly compared to the putting of extra wheels to a carriage, increasing the rattling and complexity of the machine, but adding neither to the strength of the structure nor the rapidity of its course.

SECTION IV.

1. NOTWITHSTANDING all the discussion which has taken place upon the abstract question of atmospheric motion, — and which, under certain temperatures, as we have already seen, cannot be disputed, — the true principles of ventilation still remain unsettled; and the mechanical operation of admitting the air in larger or smaller quantities with facility does not, in the slightest degree, remove the general objections that have been urged against its effects on the internal atmosphere. In considering, therefore, the question, how far the admission of external air into forcing-houses is practicable and proper, it is necessary to ask, in the first place, For what purpose is the admission of external air resorted to under certain circumstances? and, secondly, How does it act upon the atmosphere when admitted?

The first of these questions is of comparatively easy solution: the latter requires more deep consideration, and more close investigation, before we can find a satisfactory reply.

First. The necessity for ventilation arises from two prime causes, which are briefly these: to regulate and reduce the internal temperature; and to allow the escape of impure air, or that portion from which some of the essential constituents have been abstracted by the plants, or in which the natural equivalents have been changed in their proportions, and consequently the health-imbuing balance destroyed, — an effect which may arise from various causes. The first of these points is a distinct consideration, forming an important branch in vegetable physiology: the others constitute a different branch of scientific research; but in relation to our present subject, they both merge into one.

The admission of cold air as the sole or principal agent in regulating the internal temperature of a hot-house during winter, seems to be perfectly unjustifiable. There are, indeed, times when it can hardly be avoided, during the application of artificial heat; but these are exceptions, rather than the rule. Heat, when applied in early forcing, or to maintain the temperature of plant-houses, is artificial, and, therefore, so far unnatural. And it appears still more unnatural to apply more than is necessary, for the purpose of admitting the external to cool down the internal atmosphere, without having secured any equivalent advantage, but rather lost, by the change. It is much more reasonable, as well as economical, to apply *as much heat*, and *no more* than is necessary, to raise the temperature to the minimum point, or, at least, as near this point as is possible. It may be supposed that it would be unsafe to keep the temperature so close to the minimum point, lest the sudden external changes, to which we are subjected in this country in winter, might have an unfavorable effect upon the internal atmosphere; and, under certain circumstances, this would be the case,—such as an imperfect heating apparatus, a badly glazed house, or a want of skill in the management of it. The necessity of maintaining the minimum rather than the maximum temperature has been already adverted to in the preceding chapter; and, instead of being the exception to a general rule, it is rapidly becoming the rule itself. We must consider that the object to be kept in view is to improve upon the means at present in use to obtain these results, and to obviate the risk and inconvenience which might otherwise ensue by their adoption. It will be observed, that it is not when the mild and genial weather of spring is experienced that these remarks have any forcible effect, but when the outward elements are unfavorable to the development of vegetable life.

2. The atmosphere of a hot-house is very much influenced in winter by the glazing of the sashes, and the adjustment of its various parts. When the laps of the glass are open, there is a continual egress and ingress movement in the atmosphere adjacent to the apertures, extending generally over the whole of its

interior surface, but not always affecting seriously the internal volume, except in carrying off the rising particles of heated air, the greater portion of which is condensed by the cold air immediately as it escapes from the house.   The consideration which refers to the escape of air in a deteriorated state, and the consequent necessity of admitting a fresh volume in its place, does not appear to offer any insurmountable difficulties to the belief that the admission of fresh air in the months of winter is very frequently carried to an injurious excess.   Although plants, in the process of their growth, and in the discharge of their vital functions, abstract matters from the atmosphere around them, there is nothing, even in this, to render the admission of cold air in large volumes at all necessary.   In considering the nature of the atmosphere in its relation to heat and cold, its elastic and all-pervading properties must not be lost sight of.   Under any circumstances, a considerable effect will be produced by the external upon the internal atmosphere, by radiation alone ; and with the evidence before us of the successful growth of plants in situations so much closed up as in Wardian cases, we cannot do otherwise than believe that the interchange which takes place between the volumes by these causes is sufficient to secure the health and vigor of the plants, so far as the admission of air alone is concerned.   If it be argued that deterioration will take place by means of evaporation from flues, or pipes, or any substances confined within the structure, or from the decomposition of organic matter, the same fact is presented of an interchange continually going on, and is sufficient to meet the case, so far as to show, that, on this ground, at least, the admission of external air in large volumes is not essential.   Besides, with proper management, the gases that are generated by artificial heat, or by the decomposition of substances which should find a place in hot-houses, may be combined with others having an affinity for them, and thereby not only purifying the atmosphere by preventing an excess of particular agents, but also turning those agents to their legitimate purpose, and rendering them beneficial, rather than detrimental, to vegetable life.   And, therefore, it can only be in cases where misapplication or gross mismanagement of some kind or other exists, that they can possibly be productive of injury, or even of inconvenience.

These considerations, then, would seem to point out the fact, that the admission of air to any extent in forcing-houses in winter, or at a very early period of the season, cannot be said to be a matter of urgency, or necessity; neither can it be grounded on the plea that many of our practical operations have for their foundation, viz., an expedient for a better, and probably more tedious, method of effecting the same results. Whatever *impropriety* may appear in the above statement, it will be fully justified by its *truth*, — if a dozen years' extensive practice in the management of hot-houses, both large and small, and in the working of forcing-houses throughout the winter, be worth anything, as well as the evidence of many of the best practical gardeners of the present day. Then we would say that the influx of large volumes of cold air is decidedly hurtful, even on other grounds than those advanced in a former part of this chapter. But, on the other hand, the opposite extreme must also be avoided. The process may not be altogether dispensed with, although every means ought to be taken to modify its immediate effect upon the internal atmosphere. It does appear, nevertheless, that the regulation of the internal temperature, *i. e.*, the prevention of too powerful a degree of heat, when the source of that heat is the sun, is the only legitimate end to be effected by the practice. If there are any other real advantages, they are certain to follow. If air is admitted with this only in view, — and these advantages are not likely to be lost if air is not admitted when not required to effect this primary purpose, — periods of bright sunshine, then, may be regarded as the only instances in which a recourse to the practice is absolutely necessary.

3. From a full investigation and consideration of this subject, the conclusion at which we have arrived is, that, with a proper system and routine of management, as regards the application of atmospheric humidity and heat, the admission of large volumes of the external air into the interior of hot-houses is not by any means so essential as it is generally represented to be. Whatever other differences of opinion may exist with

25

respect to this practice, it cannot be denied that a risk is incurred, and frequently an injury sustained, when cold air comes in contact with the active organs of tender plants. And, therefore, if no other advantage be gained from the practice than the regulation of the temperature, then, except in cases where the heat is increased by the influence of the sun, and therefore uncontrollable, it would be a much wiser practice to apply a less amount of heat by artificial means, thus rendering it less necessary to allow the superabundant portion to escape, and consequently exposing the plants in a less degree to the risk to which we have alluded.

4. Even in those cases in which it is really necessary to have recourse to the practice of admitting air, much injury will be sustained, though it may not be apparent at the time, by admitting it in a rash and improper manner. It should be contrived so that the change to be effected may be brought about gradually, and the cold and heated volumes should be made to intermingle regularly together, and in a way that the internal volume will be equally affected by it. Thus, if it be desirable to admit a quantity of air equivalent to the reduction of 20° of temperature, then the first consideration ought to be the external temperature; and the apertures of admission ought to be regulated according to the calculations given at pp. 164 and 165, and in such a manner that the volume of air within the house will not be deteriorated thereby, nor deprived of those gases which are essential to vegetable existence.

Secondly. How does the external air act upon the internal atmosphere, when so admitted? This portion of our subject is of more difficult solution, and requires a closer investigation, inasmuch as it is influenced by various causes, such as the form of the structure, the method of admission, and the material of which the interior part of the house is composed; for example, a house presenting a large surface of glass to the morning sun requires to be sooner ventilated than one whose largest glass surface has a western aspect, and a small quantity of air admitted early in the morning will keep the temperature down for a

longer period, than a larger portion, when the temperature of the
house has increased ten or twelve degrees higher.   Again, if
the top sashes be opened first, which is generally done, then a
much larger quantity of oxygen and aqueous vapor is carried
off than at any other period of the day.   We believe it is the
practice of nineteen out of every twenty gardeners, to open the
top sashes first; then, when the internal temperature rises, and
more external air is necessary, the top sashes are opened still
more ; and, last of all, the front sashes are opened *to make a circulation;* — a circulation, indeed !   By the time the front sashes
are opened, the two atmospheres are generally equalized.   Now,
I would ask, how is this circulation produced, and what are its
effects ?   Not by the superior density of either atmosphere, for
both are the same, but by currents of wind, and draughts created
by other causes; and their effect is to carry off the moisture
already too much reduced.   The annexed figure represents a

Fig. 57.

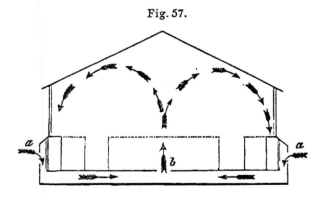

method of admitting fresh air into a house which obviates the
evil here complained of.   The air enters through the side-walls
at *a a*, then passes along beneath the floor, and enters the house
in the centre of the floor, at *b*.   In this instance, no air is
admitted at the top; hence, the air, passing through these drains,
enters the house at a higher temperature than if admitted at
the sides or top, and, becoming gradually warmed as it ascends
through the aperture in the floor, rises until it is again cooled by
action of the external air upon the glass, then falls towards both
sides of the house, producing a motion somewhat similar to that

shown by the arrows in the foregoing figure. By this method,
air may be introduced into a house at any period of the day, or
even at night; and while every advantage arising from the
admission of external air is gained, the disadvantages are done
away with, save and except by the crevices in the structure.
In winter, if cold air *must* be introduced to regulate the internal
temperature, some such method as that given above should be
adopted; but at a more advanced season of the year, when a
larger supply of air is necessary, provision must be made at the
sides for that purpose. As to opening the top sashes first, and
keeping them open till the last, it is a practice for which we are
unable to obtain any satisfactory reason, and which we think
will not bear a strict investigation. But, it may be asked, how
is the temperature to be reduced, where, at an advanced period
of spring, the sun shines more powerfully, and when the temperature of a hot-house will suddenly rise ten or fifteen degrees
above the maximum point? To answer this question, it is
necessary to consider whether there be any other method of
reducing the temperature than by expelling the heated air, by
the opening of the top sashes. From what has already been
said on this point, we think we are fully justified in disposing
of this question in the affirmative. Of course, we do not allude
to the ventilation of houses in summer, but in the months of
autumn, winter, and spring. By introducing the external air in
the manner described in the last figure, the atmosphere of a hot-
house will be reduced to any given point as effectually, though
not so rapidly, as if the heated air was expelled through the
sashes at the top of the house. This is accounted for by the
circumstance already explained, viz., that when two columns of
air of unequal temperatures are mixed together, the temperature of the whole is reduced, while its density is increased; and
hence, so long as the atmosphere continues to be heated by
reflection or radiation, this cold air will continue to cool it down,
so that nothing is lost, while all the essentials of vegetation
contained in the atmosphere are retained.

5. The materials of which the internal part of the house is
composed have also a powerful influence on the ventilation of a

hot-house. Those houses whose internal bases are composed of open soil require less ventilation than those that are paved with stone or tiles; and those that are paved with tiles, or other soft materials, require less than those formed of hard and highly reflecting bodies; dark-colored walls, also, are longer in raising the temperature of houses than walls painted white, and for this reason white is preferred to any other color, as well as for its clean and light appearance when contrasted with the dark-green foliage of the plants. But in houses that are perfectly transparent on every side, and admit abundance of light, there is no reason to suppose a dark color would not be preferable to a light one, although we are well aware that some scruples may be raised against it. Its propriety, however, can only be questioned as a matter of taste, not of utility; for, with the advantages above alluded to, in a well-constructed green-house, so far as the management of its atmosphere is concerned, we would decidedly prefer a house having the interior painted with a dark color, although we are very sensible that the effect produced would be meagre and dull, and but little calculated to harmonize with the floral inhabitants of the house, or the feelings of those who admire them.

Fig. 58.

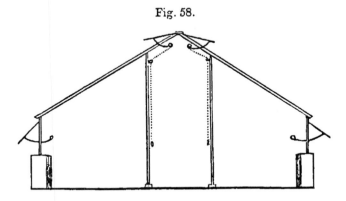

6. The above cut represents a house ventilated by the common method, i. e., the upright sashes at the sides and the top sashes along the roof, which, in span-framed houses, are generally about four feet long, or nearly square. In summer this method answers perfectly; but in winter and early spring it is

25*

next to impossible to admit air without injury to the plants, and incurring the evils which have been already detailed. Such a house as this should, by all means, have these sashes made to open, when requisite, but should also be provided with an under-ground method of admitting air, when the weather is unfavorable for opening the top and side sashes; and, in this country, this may be said to be the case for at least three months out of the twelve, during which time air can seldom be admitted in anything like a sufficient quantity, without a positive, though perhaps at the time an imperceptible, injury to exotic plants.

Various other methods have been adopted for imparting to the atmosphere of a hot-house all the freshness of the natural atmosphere, without a reduction of temperature corresponding to the amount of cold air admitted, and also to effect this without an increased consumption of fuel. The following simple method has been carried out with pretty favorable results: —

7. Suppose a house already heated by the common flue. We would propose that a square chamber be built over the top of the furnace, and embracing the neck of the flue for two or three feet, if practicable. This chamber should have a drain, not straight, but of a serpentine or zig-zag form, laid through it, one of its ends communicating with the external air, and the other communicating with the interior of the house. Into this latter opening, a pipe, made of tin or zinc, should be fitted, of sufficient size for the admission of a good volume of air. Let this pipe be laid along the lateral surface of the flue nearest the front wall of the house, not in immediate contact with, but supported by bricks, or some other means, at the distance of a few inches from the flue. Let that portion of the tube which passes along the front be perforated with holes, to facilitate the escape of the warm air, with which it will be filled, into the interior of the house. This done, let a number of small tubes, —say one for each light, or one for each alternate light, — be fixed through the front wall, or otherwise as may be convenient, one end communicating with the external atmosphere, and the other entering the perforated tube. These smaller tubes should **be**

provided with valves to open and shut at pleasure, to any extent within the limits of their diameter, so that the apertures of ingress for the cold air may be regulated by the operator according to the state of the weather and the quantity of air required. The size of these tubes will depend upon the size and situation of the house. For instance, if the house contains a large internal volume of atmosphere, the perforated tube would require to be at least eight inches in diameter, and the smaller about one half the size of the large ones. And now for its mode of action. It will be evident, that when fire is applied to the furnace, its cover (which forms the floor of the chamber) will become heated to a considerable degree. As soon as this takes place, the external valve of the drain, which communicates with the main tube, should be opened, when the external air will immediately rush in ; and, by having to traverse the heated floor of the chamber aforesaid, will expand along the large tube connected with it, which, from being in contact with the heated air, will itself become warm. The radiation of heat, too, from the surface of the flue directly beneath it, will assist in maintaining the temperature of the tube ; so that, although a portion of the heated air will escape through the perforations in its upper surface, enough will be retained to effect the purpose intended, which is, to neutralize the effects of the cold air that will be admitted through the medium of the small lateral tubes, and which may be admitted in any quantity, to the full volume of their admission. As the warm air rushes along the tube, it will mingle with that admitted by the small tubes; and the cold air, entering by the latter, will thus be modified, while a supply of fresh air will at the same time be circulated through the atmosphere of the house.

8. The advocates of what has been called a " free system of ventilation " have, like many others, in practising and advocating a favorite theory, in their excess of zeal, completely defeated the objects they sought to secure. The sole object of some of the advocates of the free system appears to be the prevention of a stagnant atmosphere. They admit an unlimited quantity of atmospheric air, at all seasons, to prevent this most terrible

evil they call stagnation, and denounce the system of sealing up plants (as some of them have termed it) from all atmospheric influence but that exerted over them by their own tainted artificial atmosphere. Now, a stagnant atmosphere, or any condition in the atmosphere of a hot-house approaching to stagnation, certainly cannot be otherwise than injurious to vegetation. This is a statement the truth of which will scarcely be called in question. But, although the prevention or removal of it has always been the chief object of every scientific gardener, it cannot be said that every gardener, having this aim in view, has taken the right way to effect his purpose; for, certainly, what is called "free" ventilation is very far from being the proper mode of obviating the evil; and, in questioning the propriety of the system upon these grounds, it may be deemed necessary to enter into an explanation of the results attributed to this system of ventilation, which is said to be requisite in order to adapt an artificial air to the circumstances of the plants growing in it, and which is supposed by some to be in exact harmony with the laws of vegetable physiology, and with all that science has unfolded to us respecting the effects of the atmosphere upon vegetable life.

The direct effects of ventilation, of any description, are twofold, mechanical and chemical. The former embraces the influence which motion possesses over the growth of plants; and this influence has never yet been accurately defined or explained — whether it be injurious or beneficial, and in what particular degree it ceases to be so. The latter comprehends the effects of the various gases, and their influence upon the vital functions of vegetable beings. To illustrate the effects of the first of these agents, viz., motion, we may refer to the circumstance that is well known, that trees trained upon a wall, in ordinary circumstances, do not grow to such size as those standing in isolated places; but their fibre is sooner matured, and also their fruit earlier, as well as larger and more saccharine. It has been asserted that wall trees do not arrive at so great an age as others standing in exposed situations, — an assertion as foundationless as it is absurd; for it is a well-known fact, that wall trees have outlived others of the same kind, planted in similar soil, and at the same period with themselves. And yet this assertion has been made the basis of an argument in favor of free ventilation. [*Experiments of Knight, in Philosophical Transactions.*]

Surely a system must be in a tottering condition when such farfetched arguments are resorted to for its support. Nor is this a solitary instance of irrelevant arguments being brought to support untenable systems, when in a sinking condition. When a plant is in a healthy and vigorous state, its sap is propelled through its various tissues by its own vital principle, aided by the combined influence of light, and heat, and moisture. And while its vital principle remains unimpaired, and these essentials of its existence unexhausted, its functions will continue in a state of activity, until some cause, known or unknown, occur to destroy them.

Let us rehearse an argument which has been advanced to overthrow the above theory. "When a plant is young and succulent, through all its parts, then all goes on very well; but when the plant becomes more matured, and its vessels less pervious to the flow of sap, *from its increased bulk, its approach to maturity and probably its deadened susceptibility to the action of light and heat*, it is evident that to prolong the existence of such a plant, a new impulse must be communicated to its sap, by a different species of agency from that which was necessary in the case of the young plant. This impulse is imparted by *motion*, and that motion is created by the winds and currents of the atmosphere."

Such is the sum and substance of an argument which involves the solution of a most important problem in vegetable physiology; and, to the merely superficial reader, it has something very plausible in its appearance, but, unfortunately, it will not stand to be strictly investigated, for then the very breezes that are brought to support it, would sweep it away. This is more especially true when the illustration is applied to the atmosphere of hot-houses, upon which point enough has been already said in this chapter, regarding the mechanical effects of currents, to render further enlargement on this subject unnecessary.

SECTION V.

1. With respect to the chemical effects of ventilation, upon an artificial atmosphere, there are two important things to be kept in view, in providing an artificial atmosphere for plants in a glazed structure; namely, the nourishment they ought to receive from it, and how to maintain it in this nutrient state.

It is needless, in this place, to enter upon the minute detail of the various substances which enter into the composition of plants, or of the various elements which combine to form the different bodies of which they are composed, — bodies, in themselves so different in their qualities, yet so identical in their formulæ, and consisting of the same elements, united together in the same proportions. This is one of those facts in chemical science which appear so very remarkable to those who have not directed their attention to chemistry, but are scarcely capable of being clearly comprehended and explained, even by those who have profoundly studied this branch of natural science. Starch and sugar — how different their properties ! — how unlike their uses ! — how unequal their importance to the human race ! Yet they consist of the same weights, of the same substances differently conjoined. The skilful architect can put together the same proportions of the same stone and cement; and the painter can combine the same colors, to produce a thousand varied impressions on the sense of sight. But in the hand of the Deity matter is infinitely more plastic. In his hands, and at his bidding, the same particles can unite in the same quantities, so as to produce the most dissimilar impressions, and on *all* our senses at once.

A knowledge of the above close relations, in composition among a class of substances occurring so abundantly in plants, imparts a degree of simplicity to our ideas of this otherwise so very complicated subject. It does not appear so mysterious that we should have woody fibre, and starch, and gum, and sugar, occurring together in variable quantities, when we know that they all are made up of the same materials, in the same proportions; or that one of these should occasionally disappear from a plant, to be replaced in whole or in part by another.

A further question arises in our minds, in connection: Are these elements formed in an artificial atmosphere, such as that of a hot-house, from the same combinations of matter as in the natural atmosphere? A reply, though probably not a satisfactory one, may be drawn from the following considerations:

During the day plants assimilate carbonic acid, and evolve oxygen; and during the night this system is reversed, although we have no accurate data from which to conclude that the relative proportions of these gases are, at all times and under all circumstances, the same. From the latest experiments, we are induced to suppose that, in an artificial atmosphere, oxygen is the most important element to be attended to, in the regulation of its elements; and from the fact that its presence, to the amount of 21 per cent. in common atmospheric air, is essential to the existence of animals and plants, there can be little doubt that it is more frequently in deficiency, than in excess, in an artificial atmosphere, and that hot-house plants are more frequently injured by the want of a proper supply, than by an excess of it in the atmosphere, when we consider the quantity of this substance which nature has stored up for the use of plants and animals. Nearly one half of the solid rocks which compose the crust of our globe, — of every solid substance we see around us, — of the houses in which we live, and of the stones on which we tread, — of the soils which we daily cultivate, — and much more than one half by weight of the bodies of all living animals and plants, — consist of this elementary body, oxygen, known to us only in the state of a gas. It may appear surprising that any one elementary substance should have been formed, by the Creator, in such abundance as to constitute nearly one half by weight of the entire crust of our globe. But this is not so surprising, when we consider that it is on the presence of this element that all animal and vegetable life depends! Nor is it less wonderful that a substance, which we know only in a state of thin air, should, by some extraordinary mechanism, by bound up and imprisoned, in such vast stores, in the solid mountains of the globe, — be destined to pervade and refresh all nature, in the form of water, and to beautify and adorn the earth in the solid parts of animals and plants. But all nature is full of similar wonders, and every step we advance in the study of the principles of our art, we cannot fail to perceive the united skill and bounty of the same great *Contriver.*

2. It has been stated by some philosophers, that when the leaves of plants are in a state of rest, their respiration is reduced to its *minimum* point, and that it increases within certain limits, as motion is communicated to them by the action of a current of air. Now this *may* be perfectly correct, and very likely *is* so; although, under natural conditions, the suspension of respiration has never been accurately ascertained. Various physiologists have attempted to discover the minimum of respiratory suspension, under certain atmospheric conditions, but without any satisfactory results. But it does not require the discovery of this delicate point, to decide on the propriety or utility of atmospheric motion. That a certain motion in the atmosphere is beneficial, we know; but then, it becomes a question of degree. We know that the gentle zephyr is favorable to vegetation, and, even in a hot-house, we have some reason to suppose it is so, *under certain circumstances,* and to a certain extent. Now it is under the *uncertain* circumstances, and the *uncertain* extent to which this practice is carried, that we have any objections; for such circumstances, and such indiscriminate abuse of the practice, we know to exist; and hence the chemical effects of ventilation, in the majority of cases, instead of promoting respiration, rather tend to prevent it, by depriving the atmosphere of the principal element that nature has designed to carry on the work.

The mechanical and chemical influences are intimately connected with each other, so that to secure the chemical advantage of ventilation, I presume consists in maintaining the proper equivalents of the atmosphere, which nature has determined as essential to the development of vegetation. If this view be correct, the grand and important practical question

suggests itself, whether, in the atmosphere of hot-houses generally, these essentials to the growth of plants be suitably provided.

By chemical research, we find that nitrogen forms only a
small portion of plants, but it is never entirely absent from any
part of them ; even when it is not found in any particular organ,
it is found to be present in the fluids that pervade it. Many
experiments have been instituted, with the view of ascertaining
expressly, by what particular organs nitrogen entered into the
plant, and in what form it enters. Indeed, this is a question
which at present occupies much attention. It is well known
that the leaves of plants absorb gaseous elements largely from
the atmosphere, both free and in a combined state, and we might,
therefore, expect that some of the nitrogen of the air would, by
this channel, be admitted into their circulation. This view,
however, is not confirmed by any of the experiments heretofore
made, with the view of investigating the action and functions of
the leaves. We are not at liberty to assume, therefore, that
any of the nitrogen which plants contain, has in this way been
derived directly from the atmosphere. It may be the case, but
it is not yet proved. There is little doubt, however, that nitrogen enters the roots of plants, in a state of solution ; but the
quantity they thus absorb is uncertain ; it is supposed to be
small, and must be variable. Therefore, by whatever organs it
finds an entrance into plants, and in whatever quantity it may
be present, the question still remains, that it is the ammonia of
the atmosphere that chiefly furnishes nitrogen to plants.

3. In a former part of this treatise, while treating on the
subject of heating, by means of fermenting manure, we have
alluded to the extraordinary effects of ammonia upon plants. It
is unnecessary, at present, to recapitulate what has already been
said on that interesting point. It has, we think, been clearly
established, that the *difference* between a hot-bed of manure, and
that heated by any other means, does not lie in the *quality* of
the heat generated ; as we know full well that a hot-bed of
manure, warmed beyond a certain point, will burn the roots
of plants as quickly as one heated by any other method to the
same temperature ; nor does it consist in any life-giving proper-

ties, possessed *ex lusively* by stable manure, for we know, also
that by placing living plants in a hot-bed, newly made, even if
the heat of the bed be kept from injuring the roots, they will
soon cease to exist as living beings, purely from an excess of
those very gases which, in proper proportions, add so much to
their natural luxuriance.   Plants are more sensitive, and more
easily affected, with regard to life and health, than many living
animals.   Many persons, who have paid little attention to vegetable physiology, may be dubious of this fact, but it is, nevertheless, true.   The atmosphere of a hot-house may be impregnated
with ammoniacal and other gases, beneficial to vegetable life,
without being offensive to the ordinary visitor, or even detected
by him in the atmosphere of the house.   Besides, it is so quickly absorbed by the plants, that it has to be saturated almost to
excess before much smell is sensibly felt.   We have carried on
the practice daily, of impregnating the atmosphere of a green-
house with carbonate of ammonia, by dissolving it in water and
sprinkling through the house, without the ammonia being detected, except by the acute olfactory organs of the experienced
chemist, except, perhaps, when the atmosphere was impregnated
to an excess, which, by way of experiment, was sometimes the
case.

4. This subject now resolves itself into the following considerations : —

(1.) Which gases is it necessary to generate artificially, for
the purpose of increasing the capacity of the atmosphere of a
hot-house to sustain vegetable life in a state of vigor and healthfulness ?

(2.) How are we to determine the precise proportions of each,
so that we may keep as near as possible to that point of healthfulness, which lies midway between deficiency and excess ?

In replying to the first question, it is not necessary to enter
into an elaborate detail of the various volatile gases which
arise from the combination of the prime elements of the organic
world, in different proportions, and which are absorbed by plants.
It may be sufficient for my present purpose, to notice that grand
stimulus of vegetation already alluded to, viz., ammonia, which,

as we have already seen, plays such an important part in the progress of vegetable life. This gas, though composed of hydrogen and nitrogen, is very unlike these, or, indeed, any other gases with which the chemist is yet acquainted. It is possessed of a most powerful penetrating smell, which is familiar to almost every one as hartshorn and smelling-salts. In excess, it suffocates living animals, though it requires a very considerable preponderance in the atmospheric volume to destroy either animal or vegetable life. Illustrations of this fact we have frequently observed in fumigating a pit, or house, for the destruction of aphides, and other insects ; but it destroys both, much more rapidly, when evolved at a high temperature, as we frequently find it in hot-beds of dung, when plants have been placed in them before the gas and heat had somewhat subsided, as well as in vineries, which we have seen filled with ammoniacal gas, when the atmosphere was near 100 degrees, when the edges of the tender leaves appeared as if they had been nipped with frost, but the insects were not entirely destroyed. In fumigating frames and pits with this and other gases, we have seen some kinds of tender-leaved plants completely destroyed, while many of the insects, tenacious of life, were uninjured, which has fully satisfied me of the truth of the statement already made, *i. e.*, that the generality of tender plants are more sensitive of noxious gases than living animals, although few may be inclined to believe it, and their disbelief is too often manifested in the treatment their plants receive. There can be little doubt that it is this gas, in a certain proportion of atmospheric air, that produces the luxuriance of plants, when combined with the mild heat of a dung-bed. Were we to ask a chemist, What are the manures which, in a fluid or gaseous state, can in these forms be presented to the atmosphere, and diffused among living plants, in a hot-house ? — he would answer, " Ammonia, obtain it from whatever source you may, either in a simple or combined state ;" and as hitherto our chief supply of this substance, which we have had to deal with in the common operations of gardening, has been found in our hot-beds of stable manure, resulting from the decomposition of vegetable matter, principally the nitrogeneous substances contained in corn and other matter on which the horses have been fed, with the compounds of salts and animal matter, all of which contain within themselves a tendency to rapid putrefaction, and necessarily evolve a large amount of ammonia. This is the principal source which gardeners have had to draw upon for a supply of this agent; and, although exercising the most striking effects, it is rather remarkable that the cause of these effects should, until lately, remain a mystery to gardeners in general, and that the same elements, in a more concentrated state, should not, in other circumstances, be applied to produce the same results.

The second question is, perhaps, of more difficult solution. Plants are living, organized, beings, and acted upon, atmospherically, chiefly by the glands that cover the surface of the leaves; and abundant evidence exists, that they are as susceptible of either injury or benefit, through the medium of the atmosphere to which they are exposed, as animal life, and our ignorance of the effect of houses artificially heated, upon the delicate organism of plants, is only accounted for from the fact, that comparatively little attention has yet been paid to this branch of horticultural science by practical gardeners, and still less has it been applied to the culture of exotic plants. If, for instance, we take a plant from the open ground, where it is fully exposed to the pure air, plant it in a pot, and place it in a close living room, or in a hot-house, the effect will be rendered obvious by the altered appearance of the plant. Again, if we take a plant newly potted, and otherwise disturbed in the roots, and set it in an arid situation, and fully exposed to the air, the leaves will be withered and dried up in a few hours, and probably the death of the plant will be the issue. But if the plants are placed in a close, *moist* atmosphere, the results will be very different. Now these illustrations are common, and, in themselves, exceedingly simple, so much so, that we frequently observe them, and, if asked the cause, we give a kind of generalizing reply, by attributing it to the sun, or some such cause, which is well known to be the principal origin of heat, yet they serve to show how susceptible plants are of influences which, strictly speaking, are neither dependent upon heat nor cold, although these two latter elements are almost the only ones which we are in the habit of supplying to our plants by measure, and that, too, in the most unnatural proportions, while the ammoniacal and hygrometrical condition of the atmosphere is generally left to uncontrolled transmutations of chance.

5. It may be asked, "What guide have we to ascertain the condition of the atmospheric gases?"

In the present state of our knowledge of gaseous bodies, their presence or preponderance in the atmosphere of hot-houses must be little else than a matter of conjecture. An experienced gardener, on entering his hot-house in the dark, can tell pretty accurately what degree of temperature the atmosphere of the house is standing at, by the sensation produced upon his face, or by the wave of his hand in the air. Now, in regard to the excess of volatile gases floating in the atmosphere, the organs of smell are much more delicate indicators than the sense of feeling. This is more especially the case when the house is close, and the temperature pretty high; for then the ammonia, being little more than half the weight of the common atmosphere, [more nearly three fifths, its specific gravity being 0.59, that of air being 1,] hence, when liberated on the floor, or on the flue, pipes, tank, or other heating apparatus, it readily rises and mingles with the atmosphere; and although it requires a considerable proportion of it in the atmosphere to be injurious, or even offensive to the senses, it is, nevertheless, easily detected by those acquainted with this gas, even when present in small quantities, and the experienced organs of the practical man have no difficulty in deciding whether or not it is present in excess. On entering a hot-house, when oxygen and aqueous vapor are deficient in the atmosphere, this fact is at once detected by the oppressive burnt smell which pervades the house. Saturate the atmosphere with water, oxygen is generated, and the smell ceases. The carbonic acid, which previously existed in excess, combines with the oxygen, and is transformed into carbonic acid gas, in which state it is assimilated by the plants. In the state of vapor, water exercises a wonderful influence over the atmosphere of a hot-house, and ministers most materially to the life and growth of plants. It is in the form of water, indeed, that nature introduces the

26*

greater portion of the oxygen which performs so important a part in the numerous and diversified changes which are continually taking place in the interior of plants. Few changes are really more wonderful, in chemical physiology, than the vast variety of transmutations which are constantly going on through the agency of the elements of water.

It rarely, perhaps never, happens that we find the same unhealthy and disagreeable smell in the external atmosphere, which we frequently perceive in forcing-houses after a strong fire has been kept up during the night. Sometimes this condition may occur in the confined streets of closely-built cities, and in the vicinity of chemical works, where the heavier gases rise into the air in a rarefied state, and, on cooling, fall again to the surface of the earth, producing sometimes injurious consequences. The combustion of fuel for the production of artificial heat produces also carbonic acid gas in great abundance. And to form this gas the oxygen is drawn from the plants to form the combination; and in this way the deficiency of oxygen, so much felt in forcing-houses, may partly be accounted for. Oxygen must exist in the atmosphere to the amount of 21 per cent. of its bulk to be capable of supporting animal and vegetable life in a state of vigorous development; and when this proportion is reduced, the plants under its influence must suffer accordingly. The most convenient method of supplying the atmosphere with oxygen is by saturation with water, which latter element contains a very large amount of this gas, — every nine pounds of this liquid containing no less than eight pounds of oxygen. In the interior of plants, water undergoes continual *de*composition and *re*composition. In its fluid state it finds its way and exists in every vessel and in every tissue; and so slight, it would appear, in such situations, is the hold which its component elements have upon each other, or so strong their tendency to combine with other substances, that they are ready to separate from each other at every impulse, yielding now oxygen to one, now hydrogen to another, as the production of the several compounds which each organ is destined to elaborate respectively demands. Yet with the same readiness do they re-attach themselves, and cling together, when new metamorphoses require it.

6. In the consti*t*ution of the natural atmosphere we are at

no loss to discover its beautiful adaptation to the wants and struct iral development of animal and vegetable life. The exciting effect of pure oxygen on the animal economy is diluted by the large admixture with nitrogen; the quantity of carbonic acid present is sufficient to supply food to the plant, while it is not so great as to prove injurious to the animal; and the watery vapor suffices to maintain the flexibility of the parts of both orders of beings, without being in such a proportion as to prove hurtful to either.*

The air, thus charged with these gases, by its subtilty diffuses itself everywhere. Into every pore of the soil it make its way. When there, it yields its oxygen, or its carbonic acid, to the dead vegetable matter existing therein, or to the living roots. When the soil is heated by the sun, the gases that are imprisoned therein expand and partially escape, and. are as before replaced by other particles of air when the heat of the sun is withdrawn.

By the action of these and other causes, a constant circulation is kept up, to a certain extent, between the atmosphere on the surface, which plays among the leaves and stems of plants, and the air which mingles with the soil and ministers to the roots;

---

* The mutual influence of animal and vegetable life is well illustrated by the following experiment. Into a glass vessel, filled with water, put a sprig of a plant and a fish. Let the vessel be tightly corked, and placed in the sun. The plant, under the influence of solar light, will soon commence the process of liberating oxygen. This being absorbed by the water is respired by the fish, which, in its turn, gives out carbonic acid to be decomposed by the plant. Remove the vessel from the sun-light; the plant will cease to give out oxygen, and the fish will soon languish, and revive when placed in the light. The moving power of this beautiful system is the solar light. The balance is thus preserved; and the atmosphere, even if of limited extent, cannot be sensibly changed through all time.

It is not intended to intimate that it is in the removal of carbonic acid from the atmosphere that plants are most essential to animals, — the supply of organic matter ready for assimilation is of more immediate importance than this, — but to show that their influence is mutually conservative, preventing that change in the constituents of the atmosphere which would *eventually* be fa al to organic life. — [*Wyman on Ventilation.*]

and will also suffice to show the absolute necessity of maintaining an adequate supply of aqueous vapor in the atmosphere of our hot-houses, as well as the imperative necessity of studying and making ourselves acquainted with the nature and qualities of the atmospheric elements. Science has already done much, and is still doing more, for the art of horticulture. We have the thermometer, by which we can deal out heat and cold by the measure. We have the barometer, by which we can ascertain to a decimal the weight or density of the air. We have, also, the hygrometer, by which we can tell the precise amount of its contained moisture, — although this latter instrument is but little used in practical horticulture, — and we hope the time is not distant when it will find a place side by side with the thermometer in our hot-houses, to which it does not yield one iota of importance, of interest, or of utility. When shall we have an instrument, equally simple and efficient as these, with which we may ascertain the proportions of its gaseous elements, so that we can regulate the constituents of an atmospheric volume as easily as we can do its heat and moisture ? Such an instrument is much wanted by exotic horticulturists, and we trust something of the kind will be yet brought into use. Such an instrument could be applied to excellent purpose, and would be an incalculable boon conferred on gardening, — one almost unequalled in importance at the present day, and would be of immense utility in all the higher and more difficult branches of exotic horticulture.

7. There is, probably, no *individual* branch of natural science so useful in itself to the practical gardener as a knowledge of the various atmospheric phenomena which occur in hot-houses, as well as out of doors; and without we study the one, we can have but little knowledge of the agencies which regulate the other. That a practical foreknowledge or intuitive perception of the ordinary changes of the atmosphere is an acquirement which may certainly be obtained, to a very considerable extent, without the aid of science, is beyond a doubt. We find that the untutored savage, taught only by his own observation, or instinctively, is regulated in his movements by an unerring perception

of the coming changes of his own peculiar climate, and many of the lower animals are also highly sensitive of changes approaching, especially the feathered tribes. Every person is more or less familiar with these facts. We reason, therefore, from the lesser to the greater; and if, in the absence of comparative calculation, or the comparison of the results of one season with another, — if, in fact, we consider what are the attainments of instinctive knowledge alone, we are justified in believing that, from established principles, the result of learned inquiry and deep investigation, and the application of science extending over many successive years, many useful facts are already known and clearly explained for our practical guidance. Aided by these researches, man's ingenuity has already turned these elements to a useful account, and made them subserve his purpose, powerful though they be. But, in rendering these powerful and all-pervading elements subservient to our will, the object of that will must be undeviatingly directed to the imitation of nature. To exceed, or even reach, in every case, the perfection of the pattern, is impossible; but the more closely it is kept in view, and the more nearly it is attained in our artificial performances, the more perfect will that performance be, and the more exactly will our own ends be answered. Any departures from the principles suggested by the examples set before us in nature, through an over-hasty desire to arrive at the object by a nearer road, not only defeats the intended purpose, but also makes the ultimate attainment of that object much more troublesome and expensive. The subject of this treatise affords too many examples of this fact; and, though these examples may remain unnoticed by some, and uncared for by others, their baneful influence on the progressing art of horticulture is neither distant nor obscure. The various structures for cultivation are, indeed, much improved of late years; so, also, are the methods of applying heat, air, vapor, and water. All are so easy, and so much improved, that we sometimes hear practical men observe, that this or that principle or system cannot be beaten or improved; yet the very best constructed apparatus, and the most perfect methods of applying heat, vapor, air, and light, are capable of astonishing improvements.　And no doubt the next twenty years will bring many a hidden treasure to light, and, in that time, even our most approved systems of applying heat, etc., will be altogether economized and reformed.

1. BEFORE concluding this brief treatise on horticultural buildings, we will just cursorily advert to one more topic connected therewith, which we are inclined to think is of far more importance than is generally credited, at least, it certainly is so, if we are to judge from the degree of its practical application, viz., the protection of plant-houses, and, more especially, forcing-houses, during cold nights, both with a view to the economizing of fuel, and the equalization of heat. If duly considered, the advantages of such covering are obvious. The low degree of night temperature, which the best cultivators of the present day agree in regarding as being most favorable to the healthfulness and general welfare of their plants, would depend upon the combustion of fuel, so much less, in proportion, as the escape of the internal heat, by radiation and otherwise, was prevented by means of a covering exterior to the conducting surface of the glass.

The manifest advantage of such a protecting body does not wholly consist in the economizing of fuel. In such a variable climate as we have in the New England States, with the external atmosphere acting on the glass at a temperature of 25 or 30 degrees below the freezing point, it is, then, almost under any system of heating, unavoidably necessary to apply an excess of artificial heat, to ensure the safety of the plants against injurious depressions of temperature. Now, if a covering of non-conducting materials be employed to intercept the action of the changing atmosphere upon the surface of the glass, the plants will be as safe at a much lower internal temperature, as if no such protection were afforded them, with a high temperature. The plants, therefore, will, under these circumstances, be in a condition more conducive to their health, than if their safety from excess of cold had involved their submission to a higher degree of artificial heat during the night.

Night coverings, moreover, seem to afford facilities for night ventilation,—a time when ventilation, of all others, appears to be most necessary; for then, deleterious gases are generated in the greatest abundance, and the agitation and circulation of the atmosphere is most required. We have seen that motion and interchange of atmospheric particles are, to a certain extent, beneficial to the health of plants; and as their functions are in a state of activity during the night, motion and circulation are as necessary during that time as at midday. If a close confined atmosphere be injurious to plants in the daytime, it must be more so during the night, especially when artificial heat is in process of generation. This fact is now beginning to be recognized by the sounds, which are echoing in our ears — though as yet but faintly, — the injunction, to *keep a little air on all night;* and which is responded to by the practice of the best cultivators of the present day.

Under ordinary circumstances, where artificial heat is necessary, there is some risk in following these recommendations. A chilly blast, which cannot be refused admission when the barrier to ingress is removed, would deal death and desolation around; and if this would be liable to happen in the daytime, when attendants are at hand, the risk would be still greater at night, when none were present to guard against it; and, under the most favorable circumstances, night ventilation, if carried to any extent, would involve a great loss of heat. It becomes, therefore, a question, if the motion and circulation of the internal atmosphere during the night could not be so far facilitated by other means, as to secure the chief advantage of an actual interchange of air, without the internal heat being carried off by the cause that produced it?

Whatever prevents the radiation of heat from the interior to the exterior atmosphere through the conducting agency of glass, decreases in the same ratio the amount of required heat, and hence, saves the plants from being subjected to unnecessary excitement. The principle upon which a covering acts efficiently, is that of enclosing a complete stratum of air between it and the glass, this body of air being entirely shut off from the surrounding outer atmosphere, as far as may be practicable to do so; and as air is a bad conductor of heat, the warmth of the interior is prevented from passing to the exterior atmosphere, by means of direct radiation from the glass; or, in other words, the exterior atmosphere, being prevented from coming in contact with the glass, cannot absorb from the interior any sensible portion of its heat. To secure this advantage, it will be evident that the covering *must* be kept some distance from the glass, and should be on every side where the structure is formed of glass; the coverings, in fact, should form a complete case to all the glazed portion of the structure.*

So far, so good. As a matter of protection, and nothing else, this is all very well. The advantages of such a covering will be obvious to every one; and, as a matter of protection alone, it deserves every word that can be said in its favor. Whether it

---

* In the different experiments, it appears that the cooling effect of wind at different velocities on a thin glass surface, is very nearly as the square root of the velocity. In these experiments, the velocity of the air was measured by the revolutions of the vanes of a fan. The temperature of the air was $68°$, the time required to cool the thermometer $20°$ was noted for every different velocity, and the maximum temperature of the thermometer in each experiment was $120°$. In still air, it equired $5' 45''$ to cool the thermometer this extent, and Table VIII. in the Appendix shows the time of cooling by air in motion.

In consequence of the large quantity of glass used in the construction of horticultural buildings, the cooling effect of wind is of considerable importance. We find, however, that, with an increased velocity, the cooling effect is considerably less in proportion, on glass, than on metal, and it will be *very* much less on window-glass than even what is stated in the table. As glass is an extremely bad conductor of heat, the increased thickness which window-glass possesses over that which forms the bulb of a thermometer, will make a material difference in the quantity of heat lost by the abduction of the air, there will be, as in this case, a greater difference between the temperature of the external and the internal surface. The cooling effect of wind, therefore, is not near so considerable as is generally supposed; and the effect of wind in hot-houses is very much increased by open laps and accidental fissures in the glazing of the sashes.

27

can practicably be made the means of admitting the external air
into the house at an increased temperature, and thereby creating a motion in the internal atmosphere, is a question which, as
yet, we are unable to prove from experience, although we mean
to take an early opportunity of testing the plan which we are
about to describe.

2.   The best material which we have seen used for this purpose is canvas, or any other kind of strong coarse cloth, painted
with two or three coats of pitch, wax, and oil boiled together, and
applied in a warm state to the cloth; this makes an efficient
and durable covering.   Asphalte felt is also used extensively
in England and Germany for this purpose.   This latter material is fixed on light wooden frames, about the size of a sash, or
larger, as may be found convenient; and for covering frames
and pits it answers admirably, as it is quite impervious to wet,
and if taken care of, will last for some years.   But for covering
the roofs of large houses, we would decidedly prefer the cloth,
which can be more easily drawn off and put on, and, if well
painted, will be as impervious to air and wet, as wooden shutters, or asphalte frames, and will be cheaper than either.

Suppose, then, that a glazed cloth, of the requisite dimensions,
is prepared.   We would provide means for securing it against
wind, by loops, etc., and fix on parallel strips of wood over each
rafter, about nine inches from the glass.   The cloth should be
made to fit quite close at the top, and to reach the ground on
all sides of the house, which, formed of conducting materials, or
side-pieces, must be made to fit closely over the over-lapping
edge of the upper one, and the lower edge secured against the
admission of air.   The house is now in a case, impervious both
to air and water, and enclosing a stratum of air, which gradually becomes warmer than the external atmosphere, and effectually prevents the latter from abstracting the heat from the interior of the house.   Then let there be square holes made along
the cloth, near the bottom, say one for each alternate light, each
aperture made about ten inches square, and provided with a
shutter of the same material, to close it when necessary.   All
these apertures, or any number of them, may be opened, accord

ing to the wind, or other circumstances likely to affect the internal atmosphere.   Then small apertures may be left open in different parts of the house, during the night, whereby an interchange of the atmospheric volume would take place, without exposing the plants to immediate contact with the cold air.   By this plan, we conceive that direct benefit would accrue to the plants, because the air between the covering and the glass, although not cold, would nevertheless be of greater density than that of the house, and would consequently find its way into the interior, by the ventilators left open for that purpose.   This would also enable us to maintain a much lower night temperature than could possibly be otherwise done, with regard to the safety of the plants, which the fear of sudden changes during the night, and consequent injury from frost, prevent from being realized in this changeable climate.

It is truly remarkable how very slight a covering is required to exclude a pretty severe frost.   " I have often," observes Dr. Wells, " in the pride of half-knowledge, smiled at the means frequently employed by gardeners to protect tender plants from cold, as it appeared to me impossible that a thin mat, or any such thin substance, could prevent them from attaining the temperature of the surrounding atmosphere, by which alone, I thought them liable to be injured.   But when I had learned that bodies on the surface of the earth, become, during a still and serene night, colder than the atmosphere, by radiating their heat to the heavens, I perceived immediately a just reason for the practice which I had before deemed useless.   Being desirous, however, of acquiring some precise information on this subject, I fixed perpendicularly in the earth of a grass plot four small sticks, and over their upper extremities, — which were six inches above the grass, and formed the corners of a square, the sides of which were two feet long, — fixed a thin cambric handkerchief, so as to cover the included space.   In this disposition of things, therefore, nothing existed to prevent the free passage of air from the surrounding grass to that which was sheltered under the handkerchief, except the four small upright sticks supporting it, and there was no substance to radiate heat downwards to the grass beneath but the cambric handkerchief.   The temperature of the grass, which was thus shielded from the sky, was, upon many nights afterwards, examined by me, and was always found higher than the neighboring grass which was uncovered, if this was colder than the air.   When the difference in temperature between the air several feet above the ground and the unsheltered grass did not exceed 5°, the sheltered grass was about as warm as the air.   If that difference, however, exceeded 5°, the air was found to be somewhat warmer than the sheltered grass. Thus, upon one night, when fully exposed grass was 11° colder than the air, the latter was 3° warmer than the sheltered grass, And the same difference existed on another night, when the air was 14° warmer than the exposed grass.   One reason for this difference, no doubt, was, that the air which passed from the exposed grass, by which it had been very much cooled, had passed through that under the handkerchief, and deprived the latter of part of its heat.   Another reason might be given, — that the handkerchief, from being made colder than the atmosphere, by the radiation of its upper surface to the heavens, would remit somewhat less to the grass beneath, than what it received from that substance.   But still, as the sheltered grass, notwithstanding these drawbacks, was, upon one night, as may be seen from the preceding account, 8°, and upon another, 11°, warmer than grass freely exposed to the sky, a sufficient reason was now obtained for the utility of a very slight covering, to protect plants from the influence of frost or external cold." *

* As the elevation of temperature, induced by the heat of summer, is essential to the full exertion of the energies of the vital principle, so the depression of temperature, consequent upon intense cold nights, has been thought to suspend the exertion of the vital energies altogether.   But this opinion is evidently founded on a mistake, as is proved by the example of such plants as protrude their leaves and flowers in the winter season only, as well as by the dissection of the yet unfolded bud, at different periods of the winter, which proves regular and progressive development; even in the case of such plants as protrude their leaves and flowers in the spring and summer, and in which, as we have said, there is a gradual, regular, and incipient development of parts, from the time of the bud's first appearance, till its ultimate opening in the spring.   The sap, it is true, flows much less freely, but it is not wholly stopped.   **Du Hamel** planted some young trees in the autumn, cutting off all the

We have instituted numerous experiments with the view of ascertaining the capacity of various substances for the protection of plants and horticultural structures, by which we find that bodies of soft and open texture, — as woollen netting, thin cloth, &c., — will, on dry, clear nights, afford an amount of protection equal to 7° of frost. But if the covering should become wet before the frost sets in, it will afford very little protection to the plants beneath it.

Coarse cloth, which had been coated with paint, kept out 10° of frost, and several kinds of plants, which, at the freezing point, would suffer injury, were kept alive during the whole winter, with the thermometer occasionally indicating 22° of frost. These plants were frequently frozen, but the covering was never removed during several months, although the air circulated freely underneath the glass.

In protecting plants, or glazed structures of any description, it is essential to observe that the covering should always be placed so that a stratum of air may always be confined between the covering and the objects to be protected; this is an important part of he matter, as, if the covering be laid immediately on the glass of a frame, or green-house, which it is wished to protect, the cold will be conducted by the covering to the glass, which in turn will cool the air beneath it. The covering should never touch the object to be sheltered, though, from what we see around us, this point appears to be very little attended to.

A covering of thin cloth, or woollen netting, when suspended in a vertical position over trees, &c., will afford better protection than the same substance laid horizontally over the surface. In this manner, wall trees are protected in the British Isles from spring frosts, and we have frequently seen the blossoms of peach, apricot, and pear trees completely uninjured under woollen or hair netting, when the hardiest trees of the woods were nipt with frost, and the tender vegetables of the garden were entirely

smaller fibers of the roots, with a view to watch the progress of the formation of new ones. At the end of a fortnight he had the plants all taken up and examined, with all possible care, to prevent injuring them, and found that, when they did not actually freeze, new roots were always uniformly developed.

27*

destroyed. Peaches and other fruit trees might frequently be protected in this way, and the crop, at least, partly saved, instead of being, in one single night, blasted for the season.

Common bass mats afford the best and cheapest protection for frames and small pits; but they have the fault of absorbing moisture very readily in wet weather, and then become very bad protection. They should never be laid on the glass in a wet state, as they are sure to do more injury than good. We have found it an excellent method, in covering frames and small houses with mats, to have a thin water-proof covering to lay over the mats, which not only prevents the escape of the confined air, but also keeps the mats always dry, and thus, one of the very best protectors is obtained.

Large structures are more difficult to cover than pits, and the difficulty which thus presents itself has, in general, prevented every attempt to overcome it. We have seen various plans put in operation, besides that which we have already described; all more or less effectual. The difficulty of getting common rollers to work in frosty weather has made them all but useless, in the protection of hot-houses by rolling blinds, or screens of oil-cloth. Nevertheless, this plan is not only an effectual one, but one which is cheap and easily adopted. And the cloth can be drawn off, in the mornings, and spread out to dry on the snow, or hung on a fence, during the day. When the time comes for covering at night, it might be so arranged as to be drawn up by cords passing through a pulley at each end of the house. We have succeeded in arrangements of this kind; and the saving of fuel in a severe winter, with the certainty of the plants being safe from injury, either from frost or from fire, is ample compensation for the trouble which it costs.

Whatever kind of object it is wished to protect, whether a house or a plant, the protector should always be at least one foot from it. A considerable difference of temperature is always observed, on still and serene nights, between bodies sheltered from the sky by substances touching them, and similar bodies which were sheltered by a substance a little above them. "I found, for example," says Dr. Wells, "upon one night, that the **warmth** of grass sheltered by a cambric handkerchief, raised a

few inches in the air, was 3° greater than a neighboring piece
of grass, which was sheltered by a similar handkerchief, which
was actually in contact with it. On another night the difference
between the temperatures of the two portions of grass, sheltered
in the same manner as the two above mentioned, from the influence of the sky, was 4°. Possibly," says he, "experience has
long ago taught gardeners the superior advantages of defending
tender plants from the cold of clear and calm nights, by means
of substances not directly touching them, though I do not recol-
lect ever having seen any contrivance for keeping mats, and such
like bodies, at a distance from the plants which they were meant
to protect." We know this to be a fact; for gardeners seldom
take any thought whether the plant is protected or not, providing it be covered, with mats or something else, from the external
atmosphere.

Straw, and corn stalks, afford good protection to trees and half
hardy shrubs, when properly arranged, so that the covering may
be water-tight. The air that lodges among the straw, and in
the interstices of the stalks, keeps the plant within, always at a
regular temperature, and prevents sudden freezing and thawing,
which prove the destruction of tender plants.

Bodies, however, capable of absorbing heat during the day,
and parting with it at night, when the temperature of the atmosphere falls, are also useful as a means of protecting plants, &c.
Among such bodies may be classed the walls of houses, which
may be regarded useful in two ways ; namely, by the mechanical shelter they afford against cold winds, and by giving out
the warmth, during the night, which they had absorbed during
the day. It appears, however, that on clear and calm nights,
those, on which plants frequently receive much injury from cold,
walls must be beneficial in another way; namely, by preventing,
in part, the loss of heat, which the plants would sustain from
radiation, if they were fully exposed to the sky. The following
experiment was made by Dr. Wells, for the purpose of determining the justness of this opinion. A cambric handkerchief
having been placed, by means of two upright sticks, perpendicularly to a grass plot, and at right angles to the course of the air,
a thermometer was laid upon the grass close to the lower edge

of the handkerchief on its windward side.   ne thermometer
thus situated was, for several nights, compared with another
lying on the same grass plat, but on a part of it fully exposed
to the sky.   On two of these nights, the air being clear and
calm, the grass close to the handkerchief was found to be four
degrees warmer than the fully exposed grass; on a third night
the difference was six degrees.   An analogous fact is mentioned by Gersten, who says that a horizontal surface is more
abundantly dewed than one which is perpendicular to the
ground.

Snow forms an excellent covering, and seems to be a provision of nature for the protection of many tender roots and plants
which would otherwise perish.   Its usefulness as a plant-pro-
tector may be disputed, from the fact of their tops being exposed
to the influence of the atmosphere, while their roots and lower
parts only are protected.   In reply to this, however, we may
observe, that it prevents the occurrence of the cold, which bodies
on the earth acquire in addition to that of the atmosphere, by
the radiation of their heat to the heavens, in still and clear
nights.   The cause, indeed, of this additional cold, does not
constantly operate, but its presence during only a few hours,
might effectually destroy plants which now pass unhurt through
the winter.   Again, as things are, while low-growing vegetable
productions are prevented, by the covering of snow, from becoming colder than the atmosphere, in consequence of their own
radiation, the parts of trees and tall shrubs which rise above the
snow are little affected by cold from this cause; for their outer-
most twigs, now that they are destitute of leaves, are much
smaller than the thermometer suspended by us in the air, which,
in this situation, seldom became more than two degrees colder
than the atmosphere.   The large branches, too, which, if fully
exposed to the sky, would become colder than the extreme parts,
are in a great degree sheltered by them, and, in the last place,
the trunks are sheltered both by the larger and smaller parts,
not to speak of the heat they derive by conduction through the
roots, from the earth kept warm by the snow. In a similar manner is partly to be explained the way in which a layer of straw

or earth preserves vegetable matters in the fields from the injurious influence of cold during severe winters. *

When frames and such places are covered with snow, it should be allowed to remain on till it melts away by the influence of the atmosphere. In like manner, trees and shrubs should never have the snow drawn from their branches, during snow storms, except where the branches are likely to be broken down by the weight of snow lying upon them. Snow is not only the best, but also the most natural, covering during the winter months.

* That the warmth of the soil acts as a protection to plants may be easily understood. A plant is penetrated in all directions by innumerable microscopic air-passages and chambers, so that there is a free communication between its extremities. It may, therefore, be conceived that, if, as necessarily happens, the air inside the plant is in motion, the effect of warming the air in the roots will be to raise the temperature of the whole individual, and the same is true of its fluids. Now, when the temperature of the soil is raised to 50° at noonday, by the force of the solar rays, it will retain a considerable part of that warmth during the night; but the temperature of the air may fall to such a degree, that the excitability of a plant would be too much and too suddenly impaired, if it acquired the coldness of the medium surrounding it. This is prevented by the warmth communicated to the general system, from the soil through the roots, so that the lowering of the temperature of the air by radiation during the night, is unable to affect plants injuriously in consequence of the antagonist force exercised by the heated soil.

SECTION VII.

GENERAL REMARKS ON THE MANAGEMENT OF THE ATMOSPHERE OF HOT-HOUSES.

1. ONE of the most prevalent errors, and one of very considerable importance, consists in reversing the natural condition of the atmosphere in regard to the artificial regulation of the temperature during the night. The artificial climate is not rendered natural by adjusting it to the heat and light of the sun. In cloudy weather, and during night, the artificial atmosphere is kept hot by fires, and by excluding the external air; while, in clear days and during sunshine, fires are left off, or allowed to decline, the external atmosphere is admitted, and the internal atmosphere is reduced to the temperature of the air without. As heat in nature is the result of the shining of the sun, it follows that when there is most light there is most heat; but the practice in managing hot-houses is generally the reverse.

"A gardener," observes Knight, "generally treats his plants as he would wish to be treated himself, and consequently, though the aggregate temperature of his house be nearly what it ought to be, its temperature during the night, relatively to that of the day, is almost always too high.

"It is very doubtful if any point in exotic horticulture is less attended to than that which is involved in this question. We are too apt to forget that plants not only have their periodical rest of winter and summer, but they have also their diurnal periods of repose. Night and its accompanying refreshments are just as necessary to them as to animals. In all nature, the temperature of night falls below that of day, and thus, the great cause of vital excitement is diminished, perspiration is stopped, and the plant parts with none of its aqueous particles, although it continues to imbibe by all its green surface as well as by its roots. The processes of assimilation are suspended. No digestion of food and conversion of it into organized matter takes place, and instead of decomposing carbonic acid by the extrication of oxygen, they part with carbonic acid, and rob the atmosphere of its oxygen, thus deteriorating the air at night. It is, therefore, most important that the temperature of glass-houses of every kind should, under all circumstances whatever, be lower during the night than the minimum temperature of the day; and this ought to take place to a greater extent than is generally imagined among practical gardeners.

"Plants, it is true, thrive well, and many species of fruit attain their greatest state of perfection in some situations within the tropics, where the temperature in the shade does not vary in the day and night more than seven or eight degrees; but in these climates the plant is exposed during the day to the full blaze of the tropical sun, and early in the night it is regularly drenched with heavy wetting dews, and, consequently, it is very differently circumstanced in the day and night, though the temperature of the air in the shade, at both periods, be very nearly the same. I suspect," continues Knight, "that a large portion of the blossoms of the cherry and other fruit trees in the forcing-house often prove abortive, because they grow in too high and too uniform a temperature. I have been led," he says, "during the last three years, to try the effects of keeping up a much higher temperature during the day than during the night. As early in the spring as I wished the blossoms of my peach trees to unfold, my house was made warm during the middle of the day, but, towards night, it was suffered to cool, and the trees were then sprinkled, by means of a large syringe, with clear water, as nearly at the temperature as that which rises from the ground as I could obtain it, and no artificial heat was given during the night, unless there appeared a prospect of frost. Under this mode of treatment, the blossoms advanced with very great vigor, and, when expanded, were of a larger size than I had ever before seen on the same varieties.

"Another ill effect of high night temperature is, that it exhausts the excitability of the tree much more rapidly than it promotes the growth, or accelerates the maturation, of the fruit, which is, in consequence, ill supplied with nutriment at the period of its ripening, when most nutriment is probably wanted. **The** Muscat of Alexandria grapes, and some other late grapes, are often seen to wither upon the branch in a very imperfect state of maturity, and the want of richness and flavor in other forced fruit is, we are very confident, often attributable to the same cause. There are few peach houses or graperies in this country in which the night temperature does not exceed, during the months of April and May, that of the warmest valleys of Jamaica, in the hottest period of the year. And there are probably as few hot-houses in which the trees are not more strongly stimulated by the close and damp air of the night, than by the temperature of the dry air of the noon of the following day. The practice which occasions this cannot be right; it is in direct opposition to nature."*

We have fully satisfied ourselves that a high night temperature is injurious to plants of any description, kept under glass, and that green-house plants not only expand their flowers more perfectly, but continue much longer in bloom, when the temperature of the house is reduced at night by the admission of air or otherwise. In like manner, fruits are not only better flavored, — a fact generally admitted, — but also better colored, and more perfect in form, by a low temperature at night. On the other hand, too much air is generally admitted during the day.

There is no doubt that gardeners frequently err in admitting the external air into their hot-houses, etc., during the day, particularly in bright weather; and this error is so common as to form a portion of regular practice. We have seen graperies and green-houses fully exposed to the parching winds of a summer day, without screen or shelter; while the plants subjected to this treatment plainly indicated, by their appearance, its injurious effects. The climate of this country is so different in respect to its atmosphere during the day, from that of Britain, we are too apt to follow the practice of that country, where this practice is also carried to too great extent.†

* Loudon's Encyclopedia of Gardening.

† The climate of the British Isles, relatively to others in the same latitude, is temperate, humid, and variable. The moderation of its temperature and its humidity are owing to its being surrounded by water,

The striking difference which is exhibited between our conservatories and green-houses in this country, and those of England, is not so much owing to the existing peculiarities of climate, as to the methods of practice adopted by the gardeners themselves in the management of the atmosphere of their houses. However costly and faultlessly a conservatory, a hot-house, or a grapery, may be constructed, the whole success of the structure depends upon the subsequent management of its atmosphere.

The imitation of warm climates in winter, for the purpose of preserving tender plants, must not be confounded with the artificial climate created in a hot-house for the purpose of forcing or accelerating foreign or native productions. As two different objects are sought for, different courses of procedure must be adopted. All that is necessary for the preservation of green-house plants, is to keep the atmosphere at night a few degrees above the freezing point; and, indeed, if a proper attention be paid to the plants, so as to avoid an excess of moisture, there is scarcely any kind of what are generally termed hot-house plants, that will not thrive well enough under similar treatment. We have often allowed our plant-houses to fall below the freezing point in very severe nights; and when long and continued frosts set in, the plant-houses should be gradually inured to bear even a few degrees of frost below 32°; and this the plants will do without injury, if they be kept in a proper condition. When the external atmosphere is dry and mild, air should be admitted freely to the green-house during winter, but closed early in the

which, being less affected by the sun than the earth, imbibes less heat in summer, and, from its fluidity, is less early cooled in winter. As the sea on the coasts of Britain never freezes, its temperature must always be above 33° or 34°; and hence, when air from the polar regions, at a much lower temperature, passes over it, that air must be in some degree heated by the radiation of the water. On the other hand, in summer, the warm currents of air from the south necessarily give out part of their heat in passing over a surface so much lower in temperature. The variable nature of its climate is chiefly owing to the unequal breadth of watery surface which surrounds it, — on one side a channel of a few leagues in breadth, on the other, the broad Atlantic Ocean. — [*Loudon's Ency. of Gard.*]

afternoon, so as to preserve a portion of the warmth generated
by the sun's rays within the house, to maintain a slight degree
of heat in the house before the heating apparatus is set to work.

The accelerating, or forcing, of the vegetables and fruits of
temperate climates into a state of premature production is somewhat different, and more difficult, than the preservation of plants
during winter. The constitutions of the various fruit-bearing
plants, as vines, &c., require atmospheres of different temperature and moisture, and their success is dependent upon many
contingent circumstances, which never occur in the mere preservation of green-house plants.

The two principal methods of accelerating fruits in hot-
houses are, by planting them permanently in borders prepared
for them, and by planting in tubs and large pots ; and keeping
a succession of plants thus prepared, every year, to supply the
places of those which had become unfruitful by the effects of
forcing and producing a heavy crop of fruit.   .

The first of these methods has long been practised, and is,
undoubtedly, the best for permanent crops, as more fruit can be
produced in a house by this method than by the potting system.
When once planted out, however, and growing under the glass,
they cannot be removed from the house, and, consequently, are
dependent upon the cultivator for the elements of consumption,
air and water. The grand effect is produced by heat, and the
great aim is to supply just as much as will harmonize with the
light afforded by the sun, and the peculiar condition under which
the plants exist. All the operations must be natural and gradual, and a good cultivator will always follow the dictates and
example of the natural world. He will never be anxious to force
things on too rapidly, — a very common error, and a frequent
cause of failure ; he will likewise be careful to guard against
sudden checks, either by a sudden decrease of temperature, or
the reverse ; but he will endeavor to continue the natural course
of vegetation uninterruptedly through foliation, inflorescence, and
fructification.

The skilful balancing of the temperature and moisture of the
air, in cultivating the different kinds of fruits in forcing-houses,
and the just adaptation of the various seasons of growth and

maturity, constitute the most complicated and difficult part of the gardener's art. There is some danger in laying down any general rules on this subject, so much depends upon the peculiarities of the kind under cultivation, and the endless train of considerations connected with the process of forcing.

The following rules, however, may be safely stated, as deserving especial attention from the gardener in charge of hot-houses :

1. Moisture is *most* required in the atmosphere by plants when they first begin to grow, and *least* when their periodical growth is completed.

2. The quantity of atmospheric moisture required by plants is, *cæteris paribus*, in inverse proportion to the distance from the equator of the countries which they naturally inhabit.

3. Plants with annual stems require more than those with ligneous stems.

4. The amount of moisture in the air most suitable to plants at rest, is in inverse proportion to the quantity of aqueous matter they, at that time, contain. Hence the dryness required in the atmosphere, by succulent plants, when at rest.

Moisture in the atmosphere, then, is absolutely necessary to all plants, when they are in a state of rapid growth, partly because it prevents the action of perspiration becoming too violent, as it always does in a high and dry atmosphere, and partly because, under such circumstances, a considerable quantity of aqueous food is absorbed from the atmosphere, in addition to that drawn from the soil by the roots.

Excessive moisture is injurious to vegetables in winter, when their digestive and decomposing powers are feeble, and evaporation from the soil should rather be intercepted than otherwise, except when the atmosphere is dried to an unhealthy degree, by the use of fire heat.

One of the causes of the Dutch method of winter-forcing is, undoubtedly, their avoiding the necessity of winter ventilation, by intercepting the excessive vapor that rises from the soil, and would otherwise mix with the air. For this purpose they interpose screens of oiled paper between the earth and the air of their houses ; and in their pits for vegetables, they cover the surface of the ground with the same oiled paper, by which means vapor is effectually intercepted, and the atmosphere preserved from excessive moisture.

The difficulty of keeping succulent plants in damp cellars, during winter, is also owing to the same cause. Moisture, without a sufficiency of light to enable plants to decompose it, quickly destroys them.

On the other hand, the difficulty of keeping up that necessary degree of humidity in the atmosphere of a dwelling room, during the summer months, is the cause of the unhealthiness of plants kept in them ; and the fact of their position being generally in the window, where there is always a current of air from without, during the day, contributes, in a great measure, to exhaust the plants of their contained moisture, and then they gradually decline. Could the atmosphere around them be kept sufficiently moist, with plenty of light, there is no reason why they should not thrive as well as in the green-house.

We have already alluded to the injurious effects of maintaining a high temperature in green-houses and conservatories during winter. If we look over the different climates of the world, we shall find, that in each there is a season of growth, and a season in which vegetation is more or less suspended, and that these periodically alternate with the same regularity as our summer and winter. I do not know that in nature there is any exception to this rule ; for even in the Tierra Templada of Mexico, where, it is said, that, at the height of 4000 to 5000 feet, there constantly reigns the genial climate of spring, which does not vary more than 8° or 9° of temperature, — intense heat and excessive cold being alike unknown, — the mean temperature varying from 68° to 70°; we cannot suppose that, even in that favored region, a season of repose is wanting; for it is difficult to conceive how plants can exist, any more than animals, in a season of incessant excitement. Indeed, it is pretty evident that these countries have periods when vegetation ceases, for Xalapa belongs to the Tierra Templada, and we know that the Ipomea purga, an inhabitant of its woods, dies down annually, like our native Convolvuli.

From what has already been said on this subject, it is evident that the natural resting of plants from growth is a most important phenomenon, of universal occurrence, and that it takes place equally in the hottest and in the coldest regions. It is, therefore, a condition necessary to the well-being of a plant, not to be overworked under any circumstances whatever; and there cannot be any good gardening where this is not attended to, in the management of plants under glass. Rest is effected in two ways; either by a very considerable lowering of temperature, or by a degree of dryness under which vegetation cannot be sustained.

In treating on the various conditions of the atmosphere, and its effects on vegetation, we have already sufficiently explained these influences; which renders it unnecessary to recapitulate them in this place. In practice we find that the effects of a very dry atmosphere are, necessarily, an inspissated state of the sap of the plant, and this, in all cases, — if not carried to an injurious extent, — leads to the formation of blossom-buds, and of fruit. This influence, however, must be controlled by the cultivator, otherwise it will lead to inevitable failure, as the sap of the plant may be so much dried up as to prevent its accumulation in sufficient quantity, in the smaller branches, to form fruit buds. It is, nevertheless. true, that a low temperature, under the influence of much light, by retarding and diminishing the expenditure of the sap in growing plants, produces nearly similar effects, and causes an early appearance of fruit.

All the operations may be very essentially influenced by these facts, when they are fully understood to the cultivator, and, by a skilful alteration of the periods of rest, we are enabled to break in upon the natural habits of plants, and to invert them so completely, that the flowers and fruits of summer may be brought to perfection at the opposite season of the year.

By carrying out these principles, we have, for several years, succeeded in fruiting grape-vines in the months of March and April, without any extraordinary degree of excitability being exercised at any period of their growth. The whole secret of success consists in preparing the plants the preceding season, by ripening their wood at an early period of the season, and exposing them to such an amount of heat and dryness as can be obtained by presenting them, unwatered, to the influence of the sun,

28*

at an early period of summer; then, after the leaves have ripened, keep them as cool as possible for some time; thus causing a sufficient accumulation of excitability by the end of October, instead of the following month of May, at which period the fruit will be ripe.

SECTION VIII.

VENTILATION WITH FANS.

In a preceding part of this work, [see Part II., Sec. V.] we have described a method of warming hot-houses practised in Germany, in which a fan is used as a means of propelling the heated air into the apartments required to be warmed, and by which the volume of air to be heated is drawn from the external atmosphere. As an auxiliary to a heating apparatus, however, the complicated arrangements of this machine, the cost of its construction, and the expense and trouble of working it, must ever continue to prevent its adoption as a method of warming horticultural buildings, however extensive they may be. But as an auxiliary of ventilation, and as a means of creating that continual motion in the air, which some cultivators so much admire, it is undoubtedly superior to all other methods.

Fans are so common as to require very little description. The kind of machine generally used for this purpose is merely a light circular kind of wheel, composed of as many vanes or blades as the size will admit. By the constant revolution of this wheel, a movement is created in the atmosphere, which causes a change in the position of the atomic particles of the atmosphere of the room in which it is at work; but does not, as some suppose, tend to its equalization.

Fans are of two kinds, and have different methods of action. The one is termed *blowing* fans ; the other, *exhausting*, or *suction* fans. In the first case, the air in the house is driven outwards from the fan, or blown away ; in the other, it is drawn towards it. It will appear evident, however, that, in applying this machine to the creation of a movement in the atmosphere of a hot-house, various requisites must be had, namely, a moving power, constantly and steadily acting, and completely under control ; and when it is to be applied to night ventilation and motion, which appears to us the most adaptable use to which it can be applied

\

in relation to any kind of horticultural structures, then a supply
of warmed air must be kept up by means of the heating apparatus, and a channel of conduction for the vitiated air to
escape by.

In places where the mechanical power for moving a fan can
be easily obtained, this machine may be turned to excellent
advantage. The question, therefore, is not as to the adaptability
of the machine, but as to the means of working it so as to bring
it within the reach of hot-house adaptation, at a cost which
would justify us in recommending it.

There are various points to be considered in relation to drawing in fresh, and expelling foul, air from a hot-house, namely, that
we must not only expel the vitiated air from the house, but we
must introduce pure air into its place ; and that pure air must
be warmed before it is introduced. We have heard and read a
good deal about this and the other method of introducing warm
air into a hot-house ; and, in theory, many of these notions are
very plausible, but when we come to apply them to practice,
they are entire failures.

The principal objects to be obtained by an efficient system of
night ventilation may be classed as follows : —

1. The expulsion of a certain quantity of vitiated air, in a
certain time, from the whole volume contained in the house ;
and, as the impure air rises by rarefaction to the upper regions
of the house, means must be provided to carry it away, without creating counter-currents, or admitting any cold air, by the
channels of conduction thus made.

2. A quantity of air must be introduced to the internal volume equal to the quantity expelled ; otherwise the remaining
internal volume will expand, by its increased temperature, and
fill the space occupied by the decreasing volume, and thus the
air becomes more vitiated than if none had escaped. The air
thus brought in must be introduced without acting in a direct
current upon the vegetable productions within the house.

3. The air thus introduced must be warmed to a certain temperature, before it enters the house. This temperature should
be regulated by the temperature at which it is desired to maintain the internal atmosphere. If the desired temperature be

50°, the air entering should not be under that temperature, but rather a few degrees above it.

4. If the house be heated by pipes laid round the side of the house, the air thus admitted should be introduced so as to pass upward, by the side of the pipes, on entering the house. This air should pass regularly and consentaneously upwards; not in sudden blasts and currents, which have always an injurious influence on the internal atmosphere.

To effect this, a hot-air chamber should be placed in connection with the heating apparatus, from which must be laid air channels, or conduction tubes, all around the house, having apertures for the egress of the air, at distances of six or eight feet apart. Within this chamber a fan might be used for drawing in the external air and driving in the warmed air through the tube. This fan might be driven by a small windmill constructed for the purpose.

When air is under the control of a moving power, it will take any direction that is desired. It will move horizontally, or vertically, either upwards or downwards, and even in both directions, at the same time.

It is essential, however, that the supply to be warmed should be drawn from the external atmosphere; and here the fan may be used to great advantage. In no case should the supply of air be drawn from the interior of the house. The vitiated air, as it passes upward, should be allowed to pass off freely into the atmosphere.

In this country, however, the fan cannot be so advantageously applied in the ventilation of horticultural buildings, as in northern Europe, and only at night, the period when ventilation is most needful. The large amount of artificial heat necessary in our New England climate, in severe nights, is more injurious to green-house plants than the excessive heat of summer. There is no impossibility, however, in producing a constant and equable motion in the atmosphere of green-houses, at night; and this may be effected by the means which we have just explained.

Fans may also be beneficially employed in producing a cooling effect in the air at the top of the house. The injurious effect of the highly-heated air in the upper regions must be obvious. We have measured the temperature of a house 45 feet in height, and have found the temperature at the floor of the house to be 38°, while the temperature of the upper stratum was 103°, showing a difference of 65°. In many other cases, we have found the temperature of the upper stratum of air in a house, above 120°, while the water cistern, at the floor of the house was covered with ice. The application of a fan may be beneficial in reducing this temperature, and expelling the foul air collected in the upper portions, at apertures lower down the house.

Various other mechanical contrivances, besides the fan, have been used for producing motion in the atmosphere of houses. Among these may be mentioned common windmills, of which we have already spoken. The windmill ventilator is a very adaptable machine, and may be constructed very simply, in connection with a hot-house, and applied in moving the atmosphere of the house, or in propelling the warmed air through the conduction tubes with greater velocity than it would acquire by its own specific gravity. The windmill, of course, is turned by the force of the wind outside the house, and is entirely dependent upon the motion of the external air, for the power it exercises over the internal atmosphere. In hot-houses, with dome-shaped roofs, it is well adapted for drawing off the highly-heated air at the top of the house, and may be made something like the screw propeller of the steamboats, and situated directly in the apex of the roof.

Pumps have also been used for drawing off the foul air from buildings, although we are not aware that they have ever been employed for ventilating hot-houses, for which they are not at all adapted.

Chimney shafts are well adapted for producing motion in the air, by the draughts. None of these methods, however, are so useful as the fan, when mechanical means are to be applied; though, for the practical purposes of ventilation, in horticultural structures, the common process of spontaneous ventilation must, in general cases, suffice; — and, therefore, the question is, as to the means of admitting the air, and the temperature at which it

,s to be admitted. The movements of the atmosphere, caused by the difference of temperature between the external and internal volumes, have been already considered; and we now leave the subject to the consideration of those who are engaged in the practical operations of exotic horticulture.

APPENDIX.

TABLE I.

TABLE of the Expansive Force of Steam, in Atmospheres, and in lbs. per square inch ; for temperatures above 212° of Fahrenheit.

N. B. The steam is supposed to be in contact with the water from which it is formed, and the water and steam to be alike in temperature.

Heat in Degrees of Fahrenheit.	Pressure.		Heat in Degrees of Fahrenheit.	Pressure.		Heat in Degrees of Fahrenheit.	Pressure.	
	Atmospheres.	lbs.		Atmospheres.	lbs.		Atmospheres.	lbs.
212	1	15	431	23	345	646	150	2250
251	2	30	436	24	360	655	160	2400
275	3	45	439	25	375	663	170	2550
294	4	60	457	30	450	671	180	2700
308	5	75	473	35	525	679	190	2850
320	6	90	487	40	600	686	200	3000
332	7	105	499	45	675	694	210	3150
342	8	120	511	50	750	700	220	3300
351	9	135	521	55	825	707	230	3450
359	10	150	531	60	900	713	240	3600
367	11	165	540	65	975	719	250	3750
374	12	180	549	70	1050	726	260	3900
381	13	195	557	75	1125	731	270	4050
387	14	210	565	80	1200	737	280	4200
393	15	225	572	85	1275	742	290	4350
399	16	240	579	90	1350	748	300	4500
404	17	255	586	95	1425	753	310	4650
409	18	270	592	100	1500	758	320	4800
414	19	285	605	110	1650	763	330	4950
418	20	300	616	120	1800	768	340	5100
423	21	315	627	130	1950	772	350	5250
427	22	330	636	140	2100			

*** The above Table is deduced from the experiments of MM. Dulong and Arago. Their calculations extend only as far as 50 atmos-

29

pheres; from thence the pressures are now calculated to 350 atmospheres by their formula, viz. : —

$$t = \frac{\sqrt[5]{e} - 1}{\cdot 7153}$$

where e represents the pressure in atmospheres, and t the temperature above 100° of Centigrade. In this equation each 100° of Centigrade is represented by unity.

In reducing these temperatures from Centigrade to Fahrenheit's scale, where the fractions amount to ·5, they have been taken as the next degree above, and all fractions below ·5 have been rejected.

TABLE II.

TABLE of the quantity of Vapor contained in Atmospheric Air, at different Temperatures, when saturated.

Temperature of Air.	Quantity of Vapor per Cubic Foot; in Grains Weight.	Temperature of Air.	Quantity of Vapor per Cubic Foot; in Grains Weight.	Temperature of Air.	Quantity of Vapor per Cubic Foot; in Grains Weight.
20°	1·52	48°	3·98	76°	9·53
22	1·64	50	4·24	78	10·16
24	1·76	52	4·52	80	10·78
26	1·90	54	4·82	82	11·49
28	2·03	56	5·13	84	12·20
30	2·25	58	5·51	86	12·91
32	2·32	60	5·83	88	13·61
34	2·48	62	6·21	90	14·42
36	2·64	64	6·60	92	15·22
38	2·82	66	7·00	94	16·11
40	3·02	68	7·43	96	17·11
42	3·24	70	7·90	98	18·20
44	3·48	72	8·40	100	19·39
46	3·73	74	8·95		

* The above Table is computed from Dr. Dalton's Experiments on the Elastic Force of Vapor.

TABLE III.

TABLE of the Expansion of Air and other Gases by Heat, when perfectly free from Vapor.

Temperature Fahrenheit's Scale.	Expansion.	Temperature Fahrenheit's Scale.	Expansion.
32°	1000	100°	1152
35	1007	110	1178
40	1021	120	1194
45	1032	130	1215
50	1043	140	1235
55	1055	150	1255
60	1066	160	1275
65	1077	170	1295
70	1089	180	1315
75	1099	190	1334
80	1110	200	1354
85	1121	210	1372
90	1132	212	1376
95	1142		

⁎ The above numbers are obtained from Dr. Dalton's experiments, which give an average of $\frac{1}{483}$ part, or ·00207 for the expansion by each degree of Fahrenheit. Gay Lussac found it to be equal to $\frac{1}{480}$ part, or ·002083 for each degree of Fahrenheit; and that the same law extends to condensable vapors when excluded from contact of the liquids which produce them.

TABLE IV.

TABLE of the Specific Gravity and Expansion of Water at different Temperatures.

Temperature, Fahrenheit's Scale.	Expansion.	Specific Gravity.	Weight of 1 Cubic Inch, in grains.	Temperature, Fahrenheit's Scale.	Expansion.	Specific Gravity.	Weight of 1 Cubic Inch, in grains.
30	·00017	·9998	252·714	121	·01236	·9878	249·677
32	·00010	·9999	252·734	124	·01319	·9870	249·473
34	·00005	·9999	252·745	127	·01403	·9861	249·265
36	·00004	·9999	252·753	130	·01490	·9853	249·053
38	·000002	·9999	252·758	133	·01578	·9844	248·836
39	·00000	1·0000	252·759	136	·01668	·9836	248·615
43	·00003	·9999	252·750	139	·01760	·9827	248·391
46	·00010	·9999	252·734	142	·01853	·9818	248·163
49	·00021	·9997	252·704	145	·01947	·9809	247·931
52	·00036	·9996	252·667	148	·02043	·9799	247·697
55	·00054	·9994	252·621	151	·02141	·9790	247·459
58	·00076	·9992	252·566	154	·02240	·9780	247·219
61	·00101	·9989	252·502	157	·02340	·9771	246·976
64	·00130	·9986	252·429	160	·02441	·9760	246·707
67	·00163	·9983	252·349	163	·02543	·9751	246·483
70	·00198	·9981	252·285	166	·02647	·9741	246·233
73	·00237	·9976	252·162	169	·02751	·9731	245·982
76	·00278	·9972	252·058	172	·02856	·9721	245·729
79	·00323	·9967	251·945	175	·02962	·9711	245·474
82	·00371	·9963	251·825	178	·03068	·9701	245·218
85	·00422	·9958	251·698	181	·03176	·9691	244·962
88	·00476	·9952	251·564	184	·03284	·9681	244·704
91	·00533	·9947	251·422	187	·03392	·9671	244·446
94	·00592	·9941	251·275	190	·03501	·9660	244·187
97	·00654	·9935	251·121	193	·03610	·9650	243·928
100	·00718	·9928	250·960	196	·03720	·9640	243·669
103	·00785	·9922	250·794	199	·03829	·9630	243·410
106	·00855	·9915	250·621	202	·03939	·9619	243·151
109	·00927	·9908	250·443	205	·04049	·9609	242·893
112	·01001	·9901	250·259	208	·04159	·9599	242·635
115	·01077	·9893	250·070	212	·04306	·9585	242·293
118	·01156	·9885	249·876				

*** In the above Table the expansions are calculated by Dr. Young's formula, $22 f^2 (1 - ·002 f)$ in ten millionths. The diminution of specific gravity is calculated by this equation: $·0000022 f^2 - ·00000000472 f^3$. In both equations, f represents the number of degrees above or below 39° of Fahrenheit. The absolute weight of a cubic inch of water, at any temperature, may be found by multiplying the weight of a cubic inch at 39°, by the specific gravity at the required temperature.

TABLE V.

TABLE of the Specific Heat, Specific Gravity, and Expansion by Heat, of different Bodies.

Barometer 30 Inches. — Thermometer 60°.

	Specific Heat.			Weight of 100 Cubic Inches.	
	Of equal Weights, by Berard and Delaroche.	Of equal Volumes, by Petit and Dulong.	Specific gravity.	Barometer 30 Inches. Thermometer 60°.	Linear Expansion by 180° of heat, from 32° to 212°.
				Grains.	
Air (atmospheric)	·2669	. .	1·000	30·519	
—— (dry) . . . . . *Apjhon*	·2767	. .	. . .	. . .	
Aqueous vapor	·8470	. .	·633	19·321*	
Azote	·2754	. .	·9722	29·65	
—— oxide of	·2369	. .	1·5277	46·596	
Carbonic acid	·2210	. .	1·5277	46·596	
—————— oxide	·2884	. .	·9722	29·65	
Hydrogen	3·2936	. .	·0694	2·118	
Olefiant gas	·4207	. .	·9722	29·65	
Oxygen	·2361	. .	1·1111	33·888	
Water	1·000	. .		——	
				Ounces.	
Water	. .	1·000	1·000	57·87	
Bismuth	. .	·0288	9·880	571·7	
Brass	. .	. .	7·824	452·77	·00186671
—— wire	. .	. .	8·396	485·87	·00193000
Cobalt	. .	·1498	8·600	497·6	. . . .
Copper	. .	·0949	8·900	515·0	·00172244
Gold	. .	·0298	19·250	1114·0	·00146606
Glass (flint)	. .	. .	2·760	159·72	·00081166
—— (tube)	. .	. .	2·520	145·83	·00087572
Iron (cast)	. .	. .	7·248	418·9	·00111111
—— (bar)	. .	·1100	7·788	450·2	·00122045
Lead	. .	·0293	11·350	656·8	·00284836
Nickel	. .	·1035	8·279	478·5	. . . .
Pewter (fine)	. .	. . .	. . .	. .	·00228300
Platinum	. .	·0314	21·470	1242·4	·00099180
Silver	. .	·0557	10·470	605·8	·00208260
Solder (lead 2 + tin 1) .	. .	. .	. .	. .	·00250800
Spelter (brass 2+zinc 1)	. .	. .			·00205809
Steel (untempered) . . .	. .	. .	7·840	453·7	·00107875
—— (yellow tempered) .	. .	. .	7·816	452·31	·00136900
Sulphur	. .	·1880	1·990	115·1	. . . .
Tellurium	. .	·0912	6·115	353·5	. . . .
Tin	. .	·0514	7·291	421·9	·00217298
Zinc	. .	·0927	7·191	416·0	·00291200

*** Air is taken as the standard for the specific gravity of the gases, and water as the standard for the solids.

* Specific gravity of steam at 212° = ·431. Weight of 100 cubic inches, 14·630 grains.

TABLE VI.

TABLE of the Effects of Heat.

	Wedgwood's Scale.	Fahrenheit's Scale.
Greatest heat observed	185	25127
Hessian crucible fused	150	20557
Cast iron thoroughly melted	150	20577
Greatest heat of a smith's forge	125	17327
" " of a plate-glass furnace	124	17197
" " of a flint-glass ditto	114	15897
Derby porcelain vitrifies	112	15637
Welding heat of iron (greatest)	95	13427
" " " " (least)	90	12777
Fine gold melts	32	5237
Fine silver melts	28	4717
Swedish copper melts	27	4587
Brass melts	21	3807
Diamond burns	14	2897
Red heat fully visible in daylight	1	1077
Iron red-hot in the twilight		884
Charcoal burns		802
Heat of a common fire		790
Iron bright-red in the dark		752
Zinc melts (680° Davy)		700
Mercury boils (Black 600°) (Secondat 644°) Petit and Dulong)		656
" " (Crichton 655°) (Irvine 672°) (Dalton) . .		660
Lowest ignition of iron in the dark		635
Lead melts (Guyton and Irvine 594°) (Crichton)		612
Steel becomes dark blue, verging on black		600
" " a full blue		560
Sulphur burns		560
Steel becomes a bright blue		550
" " purple		530
" " brown, with purple spots		510
" " brown		490
Bismuth melts		476
Steel becomes a full yellow		470
" " a pale straw color		450
Tin melts		442
Steel becomes a very faint yellow		430
Tin 3 + lead 2 + bismuth 1, melts		334
Tin and bismuth, equal parts, melts		283
Bismuth 5 + tin 3 + lead 2, melts		212
Water boils (barometer 30 in.)		212
Water freezes		32
Milk freezes		30
Vinegar freezes		28
Sea water freezes		28
Strong wine freezes		20
Quicksilver congeals		—39
Sulphuric æther congeals		—47
Natural temperature at Hudson's Bay		—51
Great artificial cold		—91

TABLE VII.

TABLE of the Quantity of Water contained in 100 feet of Pipe of different diameters.

Diameter of Pipe.	Contents of 100 Feet in length.
Inches.	Gallons.
½	·84
1	3·39
1½	7·64
2	13·58
2½	21·22
3	30·56
4	54·33
5	84·90
6	122·26

TABLE VIII.

TABLE, showing the Effects of Wind in Cooling Glass.

Velocity of the wind in miles per hour.	Time of cooling the Thermometer 20°, from 120° to 100°, Fahrenheit.		
	Observed time of cooling.	Time reduced to decimals of a minute.	Corrected time, being the inverse of the square root of the velocities, in decimals of a minute.
3·26	2′ 35″	2·58	2·58
5·18	2 10	2·16	2·04
6·54	1 55	1·91	1·82
8·86	1 40	1·66	1·56
10·90	1 30	1·50	1·41
13·36	1 15	1·25	1·27
17·97	1 5	1·08	1·10
20·45	1 0	1·00	1·03
21·54	0 55	·91	·94
27·27	0 48	·81	·88

TABLE IX.

Experiments on the Cooling Effect of Windows.[*]

These experiments were made in a wooden house, double plastered, with a space between the two plasterings; walls 6 inches thick. Heat introduced from a hot-air furnace, heated air being shut off when the room was heated to a proper temperature. Thermometer four feet from the floor. When the windows were closed, two thicknesses of blankets were fastened closely to the window-frame internally.

Three windows, equal to 33·21 square feet; walls, 531 square feet; cubic contents of room, 1930 feet, being 9 feet high, 16·5 feet long, 13 feet wide.

The room was kept as nearly as possible under the same circumstances.

	External Thermom.	Internal Thermom.	Time. h. m.	
March 19, 1843.	26	74	9 1	Weather calm, windows
	25	64	10 15	uncovered.
			74	
	22	74	11 41	Windows covered with
	18	59	2 5	blankets.
			144	
March 20.	25	74	8 8	Windows uncovered.
	24½	64	9 24	
			76	
	24	74	10 17	Windows covered with
	22	61	12 22	blankets.
			125	
March 21.	24	74	8 51	Calm, windows covered.
	19	64	10 19	
			88	
	17	74	11 26	Windows uncovered,
	16	64	12 16	calm.
			50	

The experiments were also made in other rooms, with wooden shutters internally.

The results are as follows : —

1st, room cooled 10° in 74′ = 1° in 7·4′, windows open.
"   "   15° in 144′ = 1° in 9·6′, windows closed.
2d, room cooled 10 in 76′ = 1° in 7·6′, windows open.

* Wyman on Ventilation.

2d, room cooled 13° in 125′ = 1° in 9·6′, windows closed.
3d, room cooled 10° in 88′ = 1° in 8·8′, windows closed.
  "   "   10° in 50′ = 1° in 5·0′, windows open.

Experiment with wooden shutters : —

Room cooled 10° in 93′ = 1° in 9·3′, shutters closed.
  "   "   10° in 58′ = 1° in 5·8′, shutters open.

From the above, the effect of glass is very evident, and also the advantage of curtains and shutters. We shall not attempt to form any general rule, since it could be applied correctly only under circumstances which differed very little from the above.

The preparation for covering white cotton for interior windows is composed of 4 oz. of pulverized dry white cheese, 2 oz. of white slack lime, and 4 oz. of boiled linseed oil. These three ingredients having been mixed with each other, 4 oz. of the white of eggs, and as much of the yolk, are added, and the mixture then made liquid by heating. The oil combines easily with the other ingredients, and the varnish remains pliable and quite transparent. It is applied with a brush.

TABLE X.

Weights of Watery Vapor in one Cubic Foot of Air, at Dew-points from 0° to 100° Fahrenheit.

Degrees Fahrenheit.	Grains in a foot.	Degrees Fahrenheit.	Grains in a foot.	Degrees Fahrenheit.	Grains in a foot.	Degrees Fahrenheit.	Grains in a foot.
0	0·186	26	1·915	51	4·382	76	9·523
1	0 810	27	1·986	52	4·524	77	9·813
2	0·836	28	2·054	53	4·671	78	10·111
3	0·864	29	2·125	54	4·822	79	10·417
4	0·893	30	2·197	55	4·978	80	10·732
5	0·925	31	2·273	56	5·138	81	11·055
6	0·957	32	2·350	57	5·303	82	11·388
7	0·992	33	2·430	58	5·473	83	11·729
8	1·028	34	2·513	59	5·648	84	12·079
9	1·065	35	2·598	60	5·828	85	12·439
10	1·103	36	2·686	61	6·013	86	12·808
11	1·143	37	2·776	62	6·204	87	13·185
12	1·184	38	2·870	63	6·400	88	13·577
13	1·226	39	2·966	64	6·602	89	13·977
14	1·270	40	3·066	65	6·810	90	14·387
15	1·315	41	3·168	66	7·024	91	14·809
16	1·361	42	3·274	67	7·243	92	15·241
17	1·409	43	3·382	68	7·469	93	15·684
18	1·459	44	3·495	69	7·702	94	16·140
19	1·510	45	3·610	70	7·911	95	16·607
20	1·563	46	3·729	71	8·186	96	17·086
21	1·618	47	3·851	72	8·439	97	17 577
22	1·674	48	3·979	73	8·699	98	18·081
23	1·733	49	4·109	74	8·966	99	18·598
24	1·793	50	4·244	75	9·241	100	19·129
25	1·855						

TABLE XI.

Dalton's Table of the Force of Vapor, from 32° to 80°.

Tempe-rature.	Force of va-por in inches of mercury.	Tempe-rature.	Force of va-por in inches of mercury.	Tempe-rature.	Force of va-por in inches of mercury.
32°	0·2000	49°	0·3483	65	0·6146
33	0·2066	50	0·3600	66	0·6355
34	0·2134	51	0·3735	67	0·6571
35	0·2204	52	0·3875	68	0·6794
36	0·2277	53	0·4020	69	0·7025
37	0·2352	54	0·4171	70	0·7260
38	0·2429	55	0·4327	71	0·7507
39	0·2509	56	0·4489	72	0·7762
40	0·2600	57	0·4657	73	0·8026
41	0·2686	58	0·4832	74	0·8299
42	0·2775	59	0·5012	75	0·8581
43	0·2866	60	0·5200	76	0·8873
44	0·2961	61	0·5377	77	0·9175
45	0·3059	62	0·5560	78	0·9487
46	0·3160	63	0·5749	79	0·9809
47	0·3264	64	0·5944	80	1·0120
48	0·3372				

NOTE TO TABLE XII.—(*See next page.*)

To determine the dew-point, take two thermometers, the scales of which agree, cover the bulb of one with thin muslin, and wet it with water; swing both thermometers in the air, that they may be exposed under similar circumstances, and note the height of the mercurial column in each, after it has become stationary. Ascertain the difference between the heights of the two columns. In the following table, find a number at the top corresponding to the difference of heights, and in the left hand column the degree answering to the temperature indicated by the dry bulb thermometer; the figure at the intersection of the two lines is the dew-point.

Suppose, for instance, the dry bulb indicated 70°, and the wet bulb 61°; 70 — 61 = 9, which is found at the top of the table; in the column beneath, and against 70°, is 55°, the dew-point.

TABLE XII.

Table for ascertaining Dew-point by

Temp of air	1	2	3	4	5	6	7	8	9	10	11	12	13	14
90	88·7	87·5	86·3	85·1	83·8	82·5	81·2	79·9	78·6	77·2	75·8	74·4	73·0	71·5
89	87·7	86·5	85·3	84·0	82·7	81·4	80·1	78·8	77·4	76·0	74·6	73·2	71·8	70·3
88	86·7	85·5	84·3	83·0	81·7	80·4	79·1	77·7	76·3	74·9	73·5	72·1	70·6	69·1
87	85·7	84·5	83·2	81·9	80·6	79·3	78·0	76·6	75·2	73·8	72·4	70·9	69·4	67·9
86	84·7	83·5	82·2	80·9	79·6	78·2	76·9	75·5	74·1	72·7	71·2	69·7	68·2	66·6
85	83·7	82·4	81·1	79·8	78·5	77·2	75·8	74·4	73·0	71·5	70·0	68·5	67·0	65·4
84	82·7	81·4	80·1	78·8	77·5	76·1	74·7	73·3	71·8	70·4	68·9	67·3	65·7	64·1
83	81·7	80·4	79·1	77·8	76·4	75·0	73·6	72·0	70·7	69·2	67·7	66·1	64·5	62·8
82	80·7	79·4	78·1	76·7	75·3	73·9	72·5	71·0	69·6	68·1	66·5	64·9	63·2	61·5
81	79·7	78·3	77·0	75·6	74·2	72·8	71·4	70·0	68·4	66·9	65·3	63·7	62·0	60·3
80	78·6	77·3	76·0	74·6	73·2	71·7	70·3	68·8	67·2	65·7	64·1	62·4	60·7	58·9
79	77·6	76·3	75·0	73·5	72·1	70·7	69·2	67·6	66·1	64·5	62·8	61·1	59·4	57·6
78	76·6	75·3	73·9	72·5	71·0	69·5	68·0	66·5	65·0	63·3	61·6	59·8	58·1	56·2
77	75·6	74·2	72·8	71·4	69·9	68·4	66·9	65·3	63·7	62·1	60·3	58·5	56·7	54·8
76	74·6	73·2	71·8	70·3	68·9	67·3	65·8	64·2	62·5	60·8	59·1	57·2	55·3	53·4
75	73·6	72·2	70·7	69·2	67·7	66·2	64·6	63·0	61·3	59·5	57·7	55·9	54·0	52·0
74	72·6	71·1	69·7	68·2	66·6	65·1	63·4	61·8	60·1	58·3	56·4	54·5	52·5	50·4
73	71·5	70·1	68·6	67·1	65·5	64·0	62·3	60·6	58·8	57·0	55·1	53·1	51·1	49·0
72	70·5	69·1	67·5	66·0	64·4	62·8	61·1	59·3	57·5	55·7	53·7	51·7	49·6	47·3
71	69·5	68·0	66·5	64·9	63·3	61·6	59·9	58·1	56·2	54·4	52·4	50·3	48·1	45·7
70	68·5	67·0	65·4	63·8	62·2	60·5	58·7	56·9	55·0	53·0	51·0	48·8	46·5	44·1
69	67·4	66·0	64·3	62·7	61·0	59·3	57·5	55·6	53·7	51·6	49·5	47·3	44·9	42·4
68	66·4	64·9	63·2	61·6	59·9	58·1	56·3	54·3	52·3	50·2	48·0	45·7	43·2	40·5
67	65·4	63·8	62·2	60·5	58·7	56·9	55·0	53·0	51·0	48·8	46·5	44·1	41·5	38·8
66	64·4	62·7	61·1	59·3	57·5	55·7	53·7	51·7	49·6	47·3	45·0	42·4	39·7	36·8
65	63·3	61·7	60·0	58·2	56·4	54·5	52·5	50·4	48·2	45·8	43·4	40·7	37·9	34·8
64	62·3	60·6	58·9	57·1	55·2	53·2	51·2	49·0	46·7	44·3	41·7	39·0	36·0	32·7
63	61·3	59·6	57·8	55·8	54·0	52·0	49·8	47·6	45·1	42·7	40·1	37·1	34·0	30·5
62	60·3	58·5	56·7	54·8	52·8	50·7	48·5	46·2	43·7	41·1	38·3	35·2	31·9	28·2
61	59·2	57·4	55·5	53·6	51·5	49·4	47·1	44·7	42·2	39·4	36·4	33·2	29·7	25·7
60	58·2	56·3	54·4	52·4	50·3	48·1	45·7	43·2	40·6	37·7	34·6	31·1	27·3	23·0
59	57·2	55·3	53·3	51·2	49·1	46·8	44·3	41·7	39·0	35·9	32·6	28·9	24·8	20·1
58	56·1	54·2	52·2	50·0	47·8	45·4	42·9	40·1	37·2	34·0	30·5	26·6	22·1	17·0
57	55·1	53·1	51·0	48·8	46·5	44·0	41·4	38·5	35·5	32·1	28·3	24·1	19·2	13·5
56	54·0	52·0	49·8	47·6	45·2	42·6	39·8	36·8	33·6	30·0	26·0	21·4	16·1	9·5
55	53·0	50·8	48·6	46·3	43·8	41·1	38·2	35·1	31·8	27·8	23·4	18·4	12·4	4·9
54	51·9	49·7	47·5	45·0	42·4	39·6	36·6	33·3	29·7	25·6	20·8	15·3	8·5	-0·2
53	50·9	48·6	46·2	43·8	41·1	38·1	34·8	31·3	27·4	23·0	17·8	11·5	3·7	
52	49·8	47·5	45·1	42·4	39·6	36·6	33·2	29·7	25·3	20·5	14·8	7·8	-1·4	
51	48·8	46·4	43·8	41·1	38·2	35·0	31·4	27·4	22·9	17·7	11·3	3·3		
50	47·7	45·2	42·6	39·7	36·6	33·3	29·5	25·3	20·4	14·7	7·4	-2·0		
49	46·6	44·1	41·3	38·4	35·1	31·6	27·5	23·0	17·7	11·2	2·9			
48	45·5	42·9	40·0	37·0	33·5	29·7	25·5	20·6	14·7	7·3	-2·4			
47	44·4	41·7	38·7	35·5	31·9	27·9	23·3	17·9	11·4	3·1				
46	43·4	40·5	37·4	34·0	30·1	25·7	20·8	14·8	7·4	-2·6				
45	42·2	39·3	36·1	32·5	28·4	23·9	18·5	12·0	3·6					
44	41·1	38·1	34·7	30·9	26·2	21·7	15·8	8·5	-1·2					
43	40·1	36·8	33·2	29·3	24·7	19·4	12·9	4·6	-7·0					
42	38·9	35·6	31·8	27·6	22·7	16·9	9·7	0·2						
41	37·8	34·2	30·3	25·8	20·6	14·3	6·2	-5·0						
40	36·7	33·0	28·8	23·9	18·1	11·4	2·2							

TABLE XII. — Continued.

Observations on the Wet and Dry Bulb Thermometer.

Temp. of air	15	16	17	18	19	20	21	22	23	24	25	26	27	28
90°	70·0	68·4	66·8	65·2	63·6	61·9	60·1	58·3	56·4	54·4	52·4	50·3	48·0	45·6
89	68·8	67·2	65·6	63·9	62·2	60·5	58·7	56·9	54·9	52·9	50·8	48·5	46·2	
88	67·5	65·9	64·3	62·6	60·9	59·1	57·2	55·3	53·3	51·2	49·0	46·7	44·3	
87	66·3	64·7	63·0	61·3	59·5	57·7	55·8	53·8	51·7	49·6	47·3	44·9	42·4	
86	65·0	63·3	61·6	59·9	58·1	56·2	54·2	52·2	50·1	47·8	45·5	43·0		
85	63·7	62·0	60·3	58·5	56·6	54·7	52·7	50·6	48·4	46·1	43·6	41·0		
84	62·4	60·7	59·0	57·1	55·2	53·2	51·1	48·9	46·6	44·2	41·6	38·9		
83	61·1	59·4	57·5	55·6	53·7	51·6	49·5	47·2	44·8	42·3	39·6			
82	59·8	58·0	56·1	54·2	52·1	50·0	47·8	45·4	43·0	40·3	37·4			
81	58·5	56·6	54·7	52·7	50·6	48·4	46·1	43·6	41·0	38·2	35·2			
80	57·1	55·2	53·2	51·1	49·0	46·7	44·3	41·7	39·0	36·0				
79	55·7	53·7	51·7	49·6	47·3	45·0	42·4	39·7	36·8	33·7				
78	54·2	52·2	50·1	48·0	45·6	43·1	40·4	37·6	34·6	31·2				
77	52·8	50·7	48·5	46·2	43·8	41·2	38·4	35·4	32·2					
76	51·3	49·2	46·9	44·5	42·0	39·2	36·3	33·1	29·7					
75	49·8	47·6	45·2	42·7	40·1	37·2	34·1	30·7	27·0					
74	48·2	46·0	43·5	40·8	38·1	35·0	31·7	28·1						
73	46·6	44·2	41·6	39·0	36·0	32·8	29·2	25·3						
72	45·0	42·5	39·8	36·9	33·8	30·4	26·5	22·3						
71	43·3	40·6	37·8	34·8	31·4	27·8	23·7							
70	41·5	38·7	35·8	32·5	29·0	25·0	20·5							
69	39·7	36·8	33·6	30·2	26·3	22·0	17·1							
68	37·8	34·7	31·4	27·7	23·5	18·8								
67	35·8	32·5	29·0	25·0	20·4	15·1								
66	33·7	30·2	26·4	22·1	17·0									
65	31·5	27·8	23·6	18·8	13·2									
64	29·2	25·2	20·6	15·3										
63	27·6	22·3	17·3	11·4										
62	24·0	19·3	13·7											
61	21·2	15·9	9·6											
60	18·1	12·2												
59	14·6	7·9												
58	10·8													
57	6·4													

TABLE XIII.

A Table of the Analyses of Confined Air.

Places in which the air was collected.	Oxygen in 1000 parts.	Carbonic acid in 1000 parts.	Capacity of room.	Number of persons.	Duration of closing or remaining in room.	Vol. of air for each person for length of time.	Vol. of air for each person per hour.	
	dry air.	dry air.	cub. m.		h. m.	cub. m.	cub. m.	
I. Serre de Buffon, Garden of Plants (evening.)	230·1		273·7		12			Tropical plants. Exposed to sun two thirds of day.
II. Serre de Buffon, Garden of Plants (morning.)	229·6	0·1	273·7		24			Air taken next morning at 8 o'clock. Feb. 10, 1842.
III. Amphitheatre of Chemistry, Sorbonne, before lecture.	221·3	6·5	1000·0	400?	30			Commencement of M. Dumas' lecture. Doors closed with springs.
IV. The same after lecture.	219·6	10·3	1000·0	900	1 30	1·1	0·74	End of Dumas' lecture.
V. Bedchamber (morning.)	229·4	0·4	81·0	2	8	40·5	5·0	Air collected one metre above floor in winter. Chimney in room.
VI. Ward in Notre Dame du Rosaire à la Pitié (Females.)	229·1	0·3	1058·0	51	2 30	36·0	4·0	9 A. M., 2¼ hours after closing windows.
VII. The same.	227·2	2·3	1058·0	54	9	36·0	4·0	6 A. M. Two stoves, combustion feeble during night. Air taken 1.5 metres above floor.
VIII. Flat-roofed dormitory, Salpêtrière (Incurables.)	225·2	8·0	611·1	55	6 15	11·1	1·4	Doors and windows closing badly. Air taken at 0·60 metres above floor. Odor sensible.
IX. Dormitory Salpêtrière (Epileptics.)	226·0	5·8	2217·0	121	9	19·9	2·2	Doors and windows closing better. Odor sensible. Air taken 0·60 metres.
X. Ward of Asylum (Préau.)	227·1	2·7	200·0	116*	3			Odor disagreeable. Door half open.
XI. Primary School-room, fully ventilated.	223·4	lost.	721·0	180†	4			No appreciable odor. 1080 cubic metres escaping hourly through ventilators.
XII. The same, ill ventilated.		4·7	721·0	160	4			No sensible odor. 837 cubic metres escaping hourly.
XIII. The same, entirely closed.		8·7	721·0	180	4	3·1	0·77	All apertures closed. Sensation of heat, accelerated respiration. Internal temperature, 18° C.; external, 17° C.
XIV. Chamber of Deputies, ventilating flue.		2·5	3000·0	600	2 30			No odor. 11,000 cubic metres hourly escaping through ventilator.
XV. Opéra Comique, Pit.		2·3	3000·0	1000	2 30			Air taken at stage. 80,000 cubic metres hourly escaping by ventilator.
XVI. Opéra Comique, Boxes.		4·3	5000·0	1000	2 30			Air taken at ceiling.
XVII. Stable, closed.	222·5	1·05	339·5	9‡	7 45	37·7	4·7	Doors and windows closing badly.
XVIII. Stable, ventilated.	229·2	2·2	2980·0	57‡	8	52·2	6·5	Area of section for entrance of air 3 square metres.

* Boys and girls from 3 to 16 years.   † Boys from 7 to 10 years.   ‡ 1 Horse.

TABLE XIV.

Constitution of the Atmosphere.

Dumas and Boussingault analyzed atmospheric air by fixing its oxygen on copper, which was weighed; the azote was also collected and weighed.

1000 parts of air at Paris contained by weight : —

	Oxygen.	Azote.
April 27, fair weather, ...	229.2	770·8
" " " "	229.2	770.8
" 28, " "	230.3	769.7
" " " "	230.9	769.1
" 29, " "	230.3	769.7
" " " "	230.4	769.6
May 29, rainy, ...	230.1	769.9
July 20, mid-day, rainy, ...	230.5	769.5
" 21, midnight, clear, ...	230.0	770.0
" 26, mid-day, clear, ...	230.7	769.3
Mean, ...	230.2	769.8
By volume, ...	208	792=1000

Consumption of Oxygen and Formation of Carbonic Acid.

From experiments of Dumas on himself, it appears that about twenty cubic inches were received into the lungs at each inspiration; and from fifteen to seventeen inspirations per minute. The expired air contained from three to four per cent. of carbonic acid, and had lost from four to six per cent. of oxygen. These *data*, for each day of twenty-four hours, give,

16 insp. $\times$ 20 cubic inches = 320 cubic inches expired per minute.

19,200   "   "   "   hour.

460,800   "   "   "   day

TABLE XV.

A Table of Mean Temperatures of the hottest and coldest months.

	Latitude.	Longitude.	Warmest Month	Coldest Month	Authorities.
St. Petersburgh,	59 56 N.	30 19 E.	65·66	8·60	Humboldt.
Moscow,	55 45 N.	37 32 E.	70·52	6 08	"
Melville Island,	74 47 N.	110 43 W	39·03	-35·52	Hugh Murray.
			42·41	-32·19	Ed. Phil. Journal.
Copenhagen,	55 41 N.	12 35 E.	65·66	27·14	Humboldt.
Edinburgh,	55 57 N.	3 10 W.	59·36	38·30	"
Geneva,	46 12 N.	6 8 E.	66·56	31·16	"
Vienna,	48 12 N.	16 22 E.	70·52	26 60	"
Paris,	48 50 N.	2 20 E.	65·30	36·11	"
London,	51 30 N.	0 5 W.	64·40	37·76	"
Philadelphia,	39 56 N.	75 16 W.	77·00	32·72	"
New York,	40 40 N.	73 58 W.	80·70	25·31	"
Pekin,	39 54 N.	116 27 E.	84·38	21·62	"
Milan,	45 28 N.	9 11 E.	74 66	36·14	"
Bordeaux,	44 50 N.	0 34 W.	73·01	41·00	"
Marseilles,	43 17 N.	5 22 E.	74·66	44·42	"
Rome,	41 53 N.	12 27 E.	77·00	42·26	"
Funchal,	32 37 N.	16 56 W.	75·56	64·01	"
Algiers,	36 48 N.	3 1 E.	82·76	60·0	"
Cairo,	30 2 N.	30 18 E.	85·82	56·12	
Vera Cruz,	19 11 N.	96 1 W.	81·86	71·06	"
Havanna	23 10 N.	82 13 W.	83·84	69·98	"
Cumana,	10 27 N.	65 15 W.	84·38	79·16	"
Canton,	23 10 N.	113 13 E.	84·50	57·00	Anglo-Chinese Calendar.
Macao,	22 10 N.	113 32 E.	86·00	63·50	" " "
Canaries,	23 30 N.	16 00 W.	73·90	63·70	Brande's Journal.
Lohooghat (5800 feet above the sea.)	29 23 N.	79 56 E.	69·34	43·57	Trans. Med. Phys. Soc. Calc.
Fattehpúr,	25 56 N.	80 45 E.	74 94	58·74	Gleanings in Science.
Gurrah Warrah,	23 10 N.	79 54 E.	87·45	60·23	" " "
Calcutta,	22 40 N.	88 25 E.	85·70	66·20	" " "
	-	-	86·86	70·10	Journal As. Soc.
Ava,	21 51 N.	95 98 E.	88·15	61·12	Gleanings in Science.
Bareilly,	28 23 N.	79 23 E.	91·91	56 50	" " "
Chunar,	25 9 N.	82 51 E.	90·00	58·00	Ed. Ph. Journ.
Cape of Good Hope (Feldhausen,)	34 23 S.	18 25 E.	74·27	57·43	Herschel (MSS.)
Bahamas,	26 30 N.	78 30 W.	83·52	69·07	Hon. J. C. Lees (MSS.)
Swan River,	32 00 S.	115 50 E.	78·00	54·84	Milligan.
Bermuda,	32 15 N.	64 30 W.	76·75	57·90	Col. Emmett.

TABLE XVI.

The following proportions between the Mean Temperature of the earth, as indicated by springs, and that of the atmosphere, have been collected from various sources.

Names of Places.	Authority.	Temp. of Earth.	Mean Temp. of Atmosphere.
Berlin, ...	Wahlenberg, . . .	49·28°	46·40
Carlstrom, ...	" . . .	47·30	42·03
Upsal, ...	" . . .	43·70	42·0S
Paris, ...	(Catacombs,) . . .	53·00	51·00
Charleston, ...	Volney, ...	63·00	6S·00
Philadelphia, ...	"	53·00	53·42
Virginia, ...	"	57·00	57·00
Massachusetts, ...	Dewey, ...	47·21	44·73
Vermont, ...	Volney, ...	44·00	56·00
Raith, (Scotland,)	Ferguson, ...	47·70	47·00
Gosport, (England,)	Watson, ...	52·46	51·42
Kendal, (do.)	"	47·20	47·04
Keswick, (do.)	"	46·60	48·00
Leith, (Scotland,)		47·30	48·36
South of England, ...	Rees' Cyclo, . . .	48·00	50·62
Torrid Zone, ...	Volney, ...	63·00	81·50

30*

TABLE XVII.

Showing the Specific Gravity of different kinds of timber.

	I.	II.
Box, ...	—	942
Plum-tree, ...	—	872
Hawthorn, ...	—	871
Beech, ...	852	——
Ash, ...	845	670
Yew, ...	807	744
Elm, ...	800	568
Birch, ...	——	738
Apple, ...	733	734
Pear, ...	—	732
Yoke-elm, ...	—	728
Orange-tree, ...	705	——
Walnut-tree, ...	——	660
Pine, ...	657	763
Maple, ...	——	645
Linden-tree, ...	604	559
Cypress, ...	598	——
Cedar, ...	561	——
Horse chestnut, ...	——	551
Alder, ...	——	538
White poplar, ...	529	——
Common poplar, ...	383	387
Cork, ...	240	——

*** The column I., in the above table, exhibits the specific gravity of different woods, adopted by the *Annuaire du Bureau des Longitudes* The second column contains the results obtained by M. Karmarsch.

TABLE XVIII.

Solutions for the impregnation of wood which is exposed to the atmosphere, for the purpose of preserving it from decay.

Tar.
Sulphate of Copper.
Sulphate of Zinc.
Sulphate of Iron.
Sulphate of Lime.
Sulphate of Magnesia.
Sulphate of Barytes.
Sulphate of Soda.
Alum.
Carbonate of Soda.
Carbonate of Potash.
Carbonate of Barytes.
Sulphuric Acid.
Acid of Tar, (pyroligneous acid.)
Common Salt.
Vegetable Oils.
Animal Oils.
Coal Oil, (Naphtha,)
Resins.
Quick-lime.

Glue.
Corrosive sublimate.*
Nitrate of Potash.
Arsenical Pyrites water, — (water containing arsenical acid.)
Peat Moss, (containing tannin.)
Creosote and Eupion.
Crude Acetate, or pyrolignite of iron.
Peroxide of Tin.
Oxide of Copper.
Nitrate of Copper.
Acetate of Copper.
Solution of Bitumen, in oil of turpentine.
Yellow Cromate of Potash.
Refuse Lime-water of Gas-works.
Caoutchouc, dissolved in naptha.
Drying Oil.
Beeswax, dissolved in turpentine.
Chloride of Zinc.

* Corrosive sublimate is one of the most efficient of all these antiseptic applications. It was proposed by Mr. Kyan as a preventive of dry rot, under the idea of its acting as a poison to the fungi and insects, which were the supposed cause of the disease. But this explanation of the action of corrosive sublimate is no longer tenable, as it is generally admitted that the fungi and insects are not to be considered the origin, but the result, of the dry rot. It has been suggested that its action depends on the formation of a compound of lignum, or pure woody fibre, with corrosive sublimate, which resists decomposition in circumstances where pure lignum is liable to decay. But pure lignum possesses no tendency to combine with corrosive sublimate. The action of this substance is in reality confined to the albumen, with which it unites to form an insoluble compound, not susceptible of spontaneous decomposition, and, therefore, incapable of exciting fermentation. Vegetable and animal matters, the most prone to decomposition, are completely deprived of their property of putrefaction and fermentation by the contact of corrosive sublimate. It is on this account advantageously employed as a means of preserving animal and vegetable substances. Its expensiveness in this country is a great obstacle to its extensive employment on timber used for building purposes, for fences, bridges, &c. There is scarcely any antiseptic application so effectual. By Mr. Kyan's process, the timber to be impregnated, is sawed up into planks, and soaked for seven or eight days in a solution containing one pound of corrosive sublimate to five gallons of water. The impregnation may be easily effected in an open tank; though the best way is to impregnate the timber by placing it in an air-tight box, from which the air has been exhausted as much as possible by a pump. The solution then enters the pores of wood freely, being pressed into them by a force equal to about one hundred pounds to the square inch. — *Parnell's Applied Chemistry.*

TABLE XIX.

Table showing the Heating Power of different kinds of Wood, drawn by MM. Peterson and Schödler, from the quantity of Oxygen required to burn them.

Names of Trees.	Oxygen required to burn them.
Tila Europea, lime,	140·523
Ulmus suberosa, elm,	139·408
Pinus abies, fir,	138·377
Pinus larix, larch,	138·082
Æsculus hippocastanum, horse-chestnut,	138·002
Buxus sempervirens, box,	137·315
Acer campestres, maple,	136·960
Pinus sylvestris, Scotch fir,	136·931
Pinus pinea, pitch pine,	136·886
Populus nigra, black poplar,	136·628
Pyrus communis, pear tree,	135·881
Juglans regia, walnut,	135·690
Betula alnus, alder,	133·956
Salix fragilis, willow,	133·951
Quercus robur, oak,	133·472
Pyrus malus, apple-tree,	133·340
Fraxinus excelsior, ash,	133·251
Betula alba, birch,	133.229
Prunus cerasus, cherry-tree,	133·139
Robinea pseudacacia, acacia,	132·543
Fagus sylvatica, white beach,	132·312
Prunus domestica, plum,	132·088
Fagus sylvatica, red beach,	130·834
Diospyros ebenum, ebony,	128·178

TABLE XX.

Difference in Weight of two columns of Water, each one foot high, at various Temperatures.

Difference in temperature of the two columns of water in degrees of Fahrenheit's Scale.	Difference in weight of two columns of water, contained in different sized pipes.				Difference of a column one foot high.
	1 inch d a.	2 inches dia.	3 inches dia.	4 inches dia.	per square inch.
	grs. weight.	grs. weight.	grs. weight.	grs. weight.	grs. weight.
2°	1·5	6·3	14·3	25·4	2·028
4	3·1	12·7	28·8	51·1	4·068
6	4·7	19·1	43·3	76 7	6·108 ·
8	6·4	25·6	57·9	102·5	8·160
10	8·0	32·0	72·3	128·1	10·200
12	9·6	38·5	87·0	154·1	12·264
14	11·2	45·0	101·7	180·0	14·328
16	12·8	51·4	116·3	205·9	16·392
18	14·4	57·9	131·0	231·9	18·456
20	16·1	64·5	145·7	258·0	20·532

*** It will be observed in the above table that the amount of motive power increases with the size of the pipe; for instance, the power is 4 times as great in a pipe of 4 inches diameter as in one of 2 inches. The power, however, bears exactly the same relative proportion to the resistance, or weight of water to be put in motion in all the sizes alike; for, although the motive power is 4 times as great in pipes of 4 inches diameter, as in pipes of 2 inches, the former contains 4 times as much water as the other. The power and the resistance, therefore, are relatively the same.

INDEX TO THE ILLUSTRATIONS.

PART I.

SECTION I.

PART II.

PART III.

TABLES.

INDEX.

PART I.—CONSTRUCTION.

SECTION I.

SITUATION.

SECTION II.

DESIGN.

31

SECTION V.

MATERIALS OF CONSTRUCTION.

SECTION VI.

GLASS.

SECTION VII.

FORMATION OF GARDENS.

PART II.—HEATING.

SECTION I.

PRINCIPLES OF COMBUSTION.

SECTION II.

PRINCIPLES OF HEATING HOT-HOUSES.

SECTION III.

HEATING BY HOT WATER, HOT-AIR, AND STEAM.

PART III.—VENTILATION.

SECTION I.

PRINCIPLES OF VENTILATION.

SECTION II.

EFFECTS OF VENTILATION.

SECTION III.

METHODS OF VENTILATION.

SECTION IV.

MANAGEMENT OF THE ATMOSPHERE.

SECTION V.

CHEMICAL COMBINATIONS IN THE ATMOSPHERE OF HOT-HOUSES.

SECTION VI.

PROTECTION OF PLANT—HOUSES DURING COLD NIGHTS.

SECTION VII.

GENERAL REMARKS ON THE MANAGEMENT OF THE ATMOSPHERE OF HOT—HOUSES.

SECTION VIII.

VENTILATION WITH FANS.

CATALOGUE OF BOOKS

ON

AGRICULTURE AND HORTICULTURE,

PUBLISHED BY

A. O. MOORE,

(LATE C. M. SAXTON & COMPANY,)

No. 140 FULTON STREET, NEW YORK,

SUITABLE FOR

SCHOOL, TOWN, AGRICULTURAL, AND PRIVATE LIBRARIES.

THE AMERICAN FARMER'S ENCYCLOPEDIA. - - - **$4 00**

EMBRACING ALL THE RECENT DISCOVERIES IN AGRICULTURAL CHEMistry, and the use of Mineral, Vegetable and Animal Manures, with Descriptions and Figures of American Insects injurious to Vegetation. Being a Complete Guide for the cultivation of every variety of Garden and Field Crops. Illustrated by numerous Engravings of Grasses, Grains, Animals, Implements, Insects, &c. By GOUVERNEUR EMERSON, of Pennsylvania, upon the basis of Johnson's Farmer's Encyclopedia.

DOWNING'S (A. J.) LANDSCAPE GARDENING. - - **3 50**

A TREATISE ON THE THEORY AND PRACTICE OF LANDSCAPE GARdening. Adapted to North America, with a view to the improvement of Country Residences; comprising Historical Notices and General Principles of the Art, directions for Laying out Grounds and Arranging Plantations, the Description and Cultivation of Hardy Trees, Decorative Accompaniments to the House and Grounds, the Formation of Pieces of Artificial Waters, Flower Gardens, &c., with Remarks on Rural Architecture. Elegantly Illustrated, with a Portrait of the Author. By A. J. DOWNING.

DOWNING'S (A. J.) RURAL ESSAYS, - - - - **3 00**

ON HORTICULTURE, LANDSCAPE GARDENING, RURAL ARCHITECTURE, Trees, Agriculture, Fruit, with his Letters from England. Edited, with a Memoir of the Author, by GEORGE WM. CURTIS, and a letter to his friends, by FREDERIKA BREMER, and an elegant steel Portrait of the Author.

DADD'S ANATOMY AND PHYSIOLOGY OF THE HORSE, Plain, **2 00**
Do. Do. Do. Do. Colored Plates, **4 00**

WITH ANATOMICAL AND QUESTIONAL ILLUSTRATIONS; Containing, also, a Series of Examinations on Equine Anatomy and Philosophy, with Instructions in reference to Dissection, and the mode of making Anatomical Preparations; to which is added a Glossary of Veterinary Technicalities, Toxicological Chart, and Dictionary of Veterinary Science.

DADD'S MODERN HORSE DOCTOR. - - - - **1 00**

CONTAINING PRACTICAL OBSERVATIONS ON THE CAUSES. NATURE and Treatment of Disease and Lameness of Horses, embracing the most recent and approved methods, according to an enlightened system of veterinary therapeutics, for the preservation and restoration of health. With Illustrations.

DADD'S (GEO. H.) AMERICAN CATTLE DOCTOR, - - $1 00

CONTAINING THE NECESSARY INFORMATION FOR PRESERVING THE Health and Curing the Diseases of Oxen, Cows, Sheep, and Swine, with a great variety of Original Recipes and Valuable Information in reference to Farm and Dairy management, whereby every man can be his own Cattle Doctor. The principles taught in this work are, that all Medication shall be subservient to Nature—that all Medicines must be sanative in their operation, and administered with a view of aiding the vital powers, instead of depressing, as heretofore, with the lancet or by poison. By G. H. DADD, M. D., Veterinary Practitioner.

THE DOG AND GUN, - - - - - - - 50

A FEW LOOSE CHAPTERS ON SHOOTING, among which will be found some Anecdotes and Incidents; also, instructions for Dog Breaking, and interesting letters from Sportsmen. By A BAD SHOT.

MORGAN HORSES, - - - - - - - 1 00

A PREMIUM ESSAY ON THE ORIGIN, HISTORY, AND CHARACTERISTICS of this remarkable American Breed of Horses; tracing the Pedigree from the original Justin Morgan, through the most noted of his progeny, down to the present time. With numerous portraits. To which are added hints for Breeding, Breaking, and General Use and Management of Horses, with practical Directions for training them for exhibition at Agricultural Fairs. By D. C. LINSLEY.

SORGHO AND IMPHEE, THE CHINESE AND AFRICAN SUGAR CANES, - - - - - - - 1 00

A COMPLETE TREATISE UPON THEIR ORIGIN AND VARIETIES, CULTURE and Uses, their value as a Forage Crop, and directions for making Sugar, Molasses, Alcohol, Sparkling and Still Wines, Beer, Cider, Vinegar, Paper, Starch, and Dye Stuffs. Fully Illustrated with Drawings of Approved Machinery; With an Appendix by LEONARD WRAY, of Caffraria, and a description of his patented process of crystallizing the juice of the Imphee; with the latest American experiments, including those of 1857, in the South. By HENRY S. OLCOTT. To which are added translations of valuable French Pamphlets, received from the Hon. JOHN Y. MASON, American Minister at Paris.

THE STABLE BOOK, - - - - - - - 1 00

A TREATISE ON THE MANAGEMENT OF HORSES, IN RELATION TO Stabling, Grooming, Feeding, Watering and Working, Construction of Stables, Ventilation, Appendages of Stables, Management of the Feet, and of Diseased and Defective Horses. By JOHN STEWART, Veterinary Surgeon. With Notes and Additions, adapting it to American Food and Climate. By A. B. ALLEN, Editor of the American Agriculturist.

THE HORSE'S FOOT, AND HOW TO KEEP IT SOUND, - 50

WITH CUTS, ILLUSTRATING THE ANATOMY OF THE FOOT, and containing valuable Hints on Shoeing and Stable Management, in Health and in Disease. By WILLIAM MILES.

THE FRUIT GARDEN, - - - - - - 1 25

A TREATISE, INTENDED TO EXPLAIN AND ILLUSTRATE THE PHYSIology of Fruit Trees, the Theory and Practice of all Operations connected with the Propagation, Transplanting, Pruning and Training of Orchard and Garden Trees, as Standards, Dwarfs, Pyramids, Espalier, &c. The Laying out and Arranging different kinds of Orchards and Gardens, the selection of suitable varieties for different purposes and localities, Gathering and Preserving Fruits, Treatment of Diseases, Destruction of Insects, Description and Uses of Implements, &c. Illustrated with upwards of 150 Figures, representing Different Parts of Trees, all Practical Operations, forms of Trees, Designs for Plantations, Implements, &c. By P. BARRY, of the Mount Hope Nurseries, Rochester, N. Y.

FIELD'S PEAR CULTURE, - - - - - 75

THE PEAR GARDEN: or, a Treatise on the Propagation and Cultivation of the Pear Tree, with Instructions for its Management from the Seedling to the Bearing Tree. By THOMAS W. FIELD.

BRIDGEMAN'S (THOS.) YOUNG GARDENER'S ASSISTANT, $1 50

In Three Parts, Containing Catalogues of Garden and Flower Seed, with Practical Directions under each head for the Cultivation of Culinary Vegetables and Flowers. Also directions for Cultivating Fruit Trees, the Grape Vine, &c., to which is added, a Calendar to each part, showing the work necessary to be done in the various departments each month of the year. One volume octavo.

BRIDGEMAN'S KITCHEN GARDENER'S INSTRUCTOR,   ½ Cloth,   50
     "   "   "   Cloth,   60

BRIDGEMAN'S FLORIST'S GUIDE,   -   -   -   ½ Cloth,   50
     "   "   "   -   -   Cloth,   60

BRIDGEMAN'S FRUIT CULTIVATOR'S MANUAL,   -   ½ Cloth,   50
     "   "   "   "   -   Cloth,   60

COLE'S AMERICAN FRUIT BOOK,   -   -   -   50

Containing Directions for Raising, Propagating and Managing Fruit Trees, Shrubs and Plants; with a description of the Best Varieties of Fruit. Including New and Valuable Kinds.

COLE'S AMERICAN VETERINARIAN,   -   -   -   50

Containing Diseases of Domestic Animals, their Causes, Symptoms and Remedies; with Rules for Restoring and Preserving Health by good management; also for Training and Breeding.

SCHENCK'S GARDENER'S TEXT BOOK.   -   -   -   50

Containing Directions for the Formation and Management of the Kitchen Garden, the Culture and Use of Vegetables, Fruits and Medicinal Herbs.

AMERICAN ARCHITECT,   -   -   -   -   6 00

The American Architect, Comprising original Designs of Cheap Country and Village Residences, with Details, Specifications, Plans and Directions, and an Estimate of the Cost of Each Design. By John W. Ritch, Architect. First and Second Series, 4to, bound in 1 vol.

BUIST'S (ROBERT) AMERICAN FLOWER GARDEN DIRECTORY,   1 25

Containing Practical Directions for the Culture of Plants, in the Flower-Garden, Hot-House, Green-House, Rooms or Parlor Windows, for every Month in the Year; with a Description of the Plants most desirable in each, the nature of the Soil and Situation best adapted to their Growth, the Proper Season for Transplanting, &c.; with Instructions for Erecting a Hot-House, Green-House, and Laying out a Flower Garden : the whole adapted to either Large or Small Gardens, with Instructions for Preparing the Soil, Propagating, Planting, Pruning, Training and Fruiting the Grape Vine.

THE AMERICAN BIRD FANCIER,   -   -   -   -

Considered with reference to the Breeding, Rearing. Feeding, Management and Peculiarities of Cage and House Birds. Illustrated with Engravings. By D. Jay Browne.

REEMELIN'S (CHAS.) VINE DRESSER'S MANUAL,   -   -   50

An Illustrated Treatise on Vineyards and Wine-Making, containing Full Instructions as to Location and Soil, Preparation of Ground, Selection and Propagation of Vines, the Treatment of Young Vineyards, Trimming and Training the Vines, Manures, and the Making of Wine.

DANA'S MUCK MANUAL, FOR THE USE OF FARMERS,   -   1 00

A Treatise on the Physical and Chemical Properties of Soils and Chemistry of Manures· including, also, the subject of Composts, Artificial Manures and Irrigation. A new edition, with a Chapter on Bones and Superphosphates.

CHEMICAL FIELD LECTURES FOR AGRICULTURISTS,   -   1 00

By Dr. Julius Adolphus Stockhardt, Professor in the Royal Academy of Agriculture at Tharant. Translated from the German. Edited, with notes. by Jam s E. Teschemacher.

BUIST'S (ROBERT) FAMILY KITCHEN GARDENER, - - $0 75

CONTAINING PLAIN AND ACCURATE DESCRIPTIONS OF ALL THE Different Species and Varieties of Culinary Vegetables, with their Botanical, English, French and German names, alphabetically arranged, with the Best Mode of Cultivating them in the Garden or under Glass; also Descriptions and Character of the most Select Fruits, their Management, Propagation, &c. By ROBERT BUIST, author of the "American Flower Garden Directory," &c.

DOMESTIC AND ORNAMENTAL POULTRY. Plain Plates, - 1 00
Do. Do. Do. Colored Plates, - 2 00

A TREATISE ON THE HISTORY AND MANGEMENT OF ORNAMENTAL and Domestic Poultry. By Rev. EDMUND SAUL DIXON, A.M., with large additions by J. J. KERR, M.D. Illustrated with sixty-five Original Portraits, engraved expressly for this work. Fourth edition revised.

HOW TO BUILD AND VENTILATE HOT-HOUSES, - - 1 25

A PRACTICAL TREATISE ON THE CONSTRUCTION, HEATING AND Ventilation of Hot-Houses, including Conservatories, Green-Houses, Graperies and other kinds of Horticultural Structures, with Practical Directions for their Management, in regard to Light, Heat and Air. Illustrated with numerous engravings. By P. B. LEUCHARS, Garden Architect.

CHORLTON'S GRAPE-GROWER'S GUIDE, - - - - 60

INTENDED ESPECIALLY FOR THE AMERICAN CLIMATE. Being a Practical Treatise on the Cultivation of the Grape Vine in each department of Hot House, Cold Grapery, Retarding House and Out-door Culture. With Plans for the Construction of the Requisite Buildings, and giving the best methods for Heating the same. Every department being fully Illustrated. By WILLIAM CHORLTON.

NORTON'S (JOHN P.) ELEMENTS OF SCIENTIFIC AGRICULTURE, 60

OR, THE CONNECTION BETWEEN SCIENCE AND THE ART OF PRACTICAL Farming. Prize Essay of the New York State Agricultural Society. By JOHN P. NORTON, M.A., Professor of Scientific Agriculture in Yale College. Adapted to the use of Schools.

JOHNSTON'S (J. F. W.) CATECHISM OF AGRICULTURAL CHEM-
ISTRY AND GEOLOGY, - - - - - 25

By JAMES F. W. JOHNSTON. M.A.. F.R.SS.L. and E., Honorary Member of the Royal Agricultural Society of England, and author of "Lectures on Agricultural Chemistry and Geology." With an Introduction by JOHN PITKIN NORTON, M.A., late Professor of Scientific Agriculture in Yale College. With notes and additions by the author, prepared expressly for this edition, and an Appendix compiled by the Superintendent of Education in Nova Scotia. Adapted to the use of Schools.

JOHNSTON'S (J. F. W.) ELEMENTS OF AGRICULTURAL CHEM-
ISTRY AND GEOLOGY, - - - - - 1 00

With a Complete Analytical and Alphabetical Index and an American Preface. By Hon. SIMON BROWN, Editor of the "New England Farmer.'

JOHNSTON'S (JAMES F. W.) AGRICULTURAL CHEMISTRY, 1 25

LECTURES ON THE APPLICATION OF CHEMISTRY AND GEOLOGY TO Agriculture. New edition, with an Appendix, containing the Author's Experiments in Practical Agriculture.

THE COMPLETE FARMER AND AMERICAN GARDENER, 1 25

RURAL ECONOMIST AND NEW AMERICAN GARDENER; Containing a Compendious Epitome of the most Important Branches of Agriculture and Rural Economy; with Practical Directions on the Cultivation of Fruits and Vegetables, including Landscape and Ornamental Gardening. By THOMAS G. FESSENDEN. 2 vols. in one.

FESSENDEN'S (T. G.) AMERICAN KITCHEN GARDENER, - 50

CONTAINING DIRECTIONS FOR THE CULTIVATION OF VEGETABLES AND Garden Fruits Cloth.

NASH'S (J. A.) PROGRESSIVE FARMER, - - - - £0 60

A SCIENTIFIC TREATISE ON AGRICULTURAL CHEMISTRY, THE GE-
ology of Agriculture, on Plants and Animals, Manures and Soils, applied to Practical
Agriculture ; with a Catechism of Scientific and Practical Agriculture. By J. A. NASH

BRECK'S BOOK OF FLOWERS, - - - - - 1 00

IN WHICH ARE DESCRIBED ALL THE VARIOUS HARDY HERBACEOUS
Perennials, Annuals, Shrubs, Plants and Evergreen Trees, with Directions for their
Cultivation.

SMITH'S (C. H. J.) LANDSCAPE GARDENING, PARKS AND PLEASURE
GROUNDS, - - - - - - - - 1 25

WITH PRACTICAL NOTES ON COUNTRY RESIDENCES, VILLAS, PUBLIC
Parks and Gardens. By CHARLES H. J. SMITH, Landscape Gardener and Garden
Architect, &c. With Notes and Additions by LEWIS F. ALLEN, author of "Rural
Architecture."

THE COTTON PLANTER'S MANUAL, - - - - 1 00

BEING A COMPILATION OF FACTS FROM THE BEST AUTHORITIES ON
the Culture of Cotton, its Natural History, Chemical Analysis, Trade and Consumption,
and embracing a History of Cotton and the Cotton Gin. By J. A. TURNER.

COBBETT'S AMERICAN GARDENER, - - - - 50

A TREATISE ON THE SITUATION, SOIL, AND LAYING-OUT OF GARDENS,
and the making and managing of Hot-Beds and Green-Houses, and on the Propagation
and Cultivation of the several sorts of Vegetables, Herbs, Fruits and Flowers.

ALLEN (J. FISK) ON THE CULTURE OF THE GRAPE, - 1 00

A PRACTICAL TREATISE ON THE CULTURE AND TREATMENT OF THE
Grape Vine, embracing its History, with Directions for its Treatment in the United
States of America, in the Open Air and under Glass Structures, with and without
Artificial Heat. By J. FISK ALLEN.

ALLEN'S (R. L.) DISEASES OF DOMESTIC ANIMALS, - 75

BEING A HISTORY AND DESCRIPTION OF THE HORSE, MULE, CATTLE,
Sheep, Swine, Poultry, and Farm Dogs, with Directions for their Management, Breed-
ing, Crossing, Rearing, Feeding, and Preparation for a Profitable Market ; also, their
Diseases and Remedies, together with full Directions for the Management of the Dairy,
and the comparative Economy and Advantages of Working Animals, the Horse, Mule,
Oxen, &c. By R. L. ALLEN.

ALLEN'S (R. L.) AMERICAN FARM BOOK, - - - 1 00

THE AMERICAN FARM BOOK ; or, a Compend of American Agricul-
ture, being a Practical Treatise on Soils, Manures, Draining, Irrigation, Grasses, Grain,
Roots, Fruits, Cotton, Tobacco, Sugar Cane, Rice, and every Staple Product of the
United States ; with the Best Methods of Planting, Cultivating and Preparation for
Market. Illustrated with more than 100 engravings. By R. L. ALLEN.

ALLEN'S (L. F.) RURAL ARCHITECTURE ; - - - 1 25

BEING A COMPLETE DESCRIPTION OF FARM HOUSES, COTTAGES, AND
Out Buildings, comprising Wood Houses, Workshops, Tool Houses, Carriage and
Wagon Houses, Stables, Smoke and Ash Houses, Ice Houses. Apiaries or Bee Houses,
Poultry Houses, Rabbitry, Dovecote, Piggery, Barns, and Sheds for Cattle, &c., &c,
together with Lawns, Pleasure Grounds, and Parks ; the Flower, Fruit, and Vege-
table Garden ; also useful and ornamental domestic Animals for the Country Resident,
&c., &c Also, the best method of conducting water into Cattle Yards and Houses.
Beautifully illustrated.

WARING'S ELEMENTS OF AGRICULTURE ; - - - 75

A BOOK FOR YOUNG FARMERS, WITH QUESTIONS FOR THE USE OF
Schools

PARDEE (R. G.) ON STRAWBERRY CULTURE ; - - $0 60

A COMPLETE MANUAL FOR THE CULTIVATION OF THE STRAWBERRY ; with a description of the best varieties.

Also, notices of the Raspberry, Blackberry, Currant, Gooseberry, and Grape; with directions for their cultivation, and the selection of the best varieties. "Every process here recommended has been proved, the plans of others tried, and the result is here given." With a valuable appendix, containing the observations and experience of some of the most successful cultivators of these fruits in our country.

GUENON ON MILCH COWS ; - - -   60

A TREATISE ON MILCH COWS, whereby the Quality and Quantity of Milk which any Cow will give may be accurately determined by observing Natura' Marks or External Indications alone; the length of time she will continue to give Milk, &c., &c. By M. FRANCIS GUENON, of Libourne, France. Translated by NICHO-LAS P. TRIST, Esq ; with Int oduction, Remarks, and Observations on the Cow and the Dairy, by JOHN S. SKINNER. Illustrated with numerous engravings. Neatly done up in paper covers, 37 cts.

AMERICAN POULTRY YARD ; - - -   1 00

COMPRISING THE ORIGIN, HISTORY AND DESCRIPTION of the different Breeds of Domestic Poultry, with complete directions for their Breeding, Crossing, Rearing, Fattening, and Preparation for Market ; including specific directions for Caponizing Fowls, and for the Treatment of the Principal Diseases to which they are subject, drawn from authentic sources and personal observation. Illustrated with numerous engravings. By D. J. BROWNE.

BROWNE'S (D. JAY) FIELD BOOK OF MANURES ; - -   1 25

OR, AMERICAN MUCK BOOK : Treating of the Nature, Properties, Sources, History, and Operations of all the Principal Fertilizers and Manures in Common Use, with specific directions for their Preservation, and Application to the Soil and to Crops ; drawn from authentic sources, actual experience, and personal observation, as combined with the Leading Principles of Practical and Scientific Agriculture. By D. JAY BROWNE.

RANDALL'S (H. S.) SHEEP HUSBANDRY ; - -   1 25

WITH AN ACCOUNT OF THE DIFFERENT BREEDS, and general directions in regard to Summer and Winter Management, Breeding, and the Treatment of Diseases, with Portraits and other Engravings. By HENRY S. RANDALL.

THE SHEPHERD'S OWN BOOK ; · - - -   2 00

WITH AN ACCOUNT OF THE DIFFERENT BREEDS, DISEASES AND MANagement of Sheep, and General Directions in regard to Summer and Winter Management, Breeding, and the Treatment of Diseases; with Illustrative Engravings, by YOUATT & RANDALL; embracing Skinner's Notes on the Breed and Management of Sheep in the United States, and on the Culture of Fine Wool.

YOUATT ON SHEEP , - - - -   75

THEIR BREED, MANAGEMENT AND DISEASES, with Illustrative Engravings; to which are added Remarks on the Breeds and Management of Sheep in the United States, and on the Culture of Fine Wool in Silesia. By WILLIAM YOUATT.

YOUATT AND MARTIN ON CATTLE ; - - -   1 25

BEING A TREATISE ON THEIR BREEDS, MANAGEMENT, AND DISEASES, comprising a full History of the Various Races ; their Origin, Breeding, and Merits; their capacity for Beef and Milk. By W. YOUATT and W. C. L. MARTIN. The whole forming a Complete Guide for the Farmer, the Amateur, and the Veterinary Surgeon, with 100 Illustrations. Edited by AMBROSE STEVENS.

YOUATT ON THE HORSE ; - - - - -   1 25

YOUATT ON THE STRUCTURE AND DISEASES OF THE HORSE, with their Remedies. Also, Practical Rules for Buyers, Breeders, Smiths, &c. Edited by W. C. SPOONER, M.R.O V S. With an account of the Breeds in the United States, by HENRY S. RANDALL.

YOUATT AND MARTIN ON THE HOG; - - - - $0 75

A Treatise on the Breeds. Management, and Medical Treatment of Swine, with Directions for Salting Pork, and Curing Bacon and Hams. By Wm Youatt, V.S. and W. C. L. Martin. Edited by Ambrose Stevens. Illustrated with Engravings drawn from life

BLAKE'S (REV. JOHN L.) FARMER AT HOME; - 1 25

A Family Text Book for the Country; being a Cyclopedia of Agricultural Implements and Productions, and of the more important topics in Domestic Economy, Science, and Literature, adapted to Rural Life. By Rev. John L. Blake, D. D.

MUNN'S (B.) PRACTICAL LAND DRAINER; - - - 50

Being a Treatise on Draining Land, in which the most approved systems of Drainage are explained, and their differences and comparative merits discussed; with full Directions for the Cutting and Making of Drains, with Remarks upon the various materials of which they may be constructed. With many illustrations. By B. Munn, Landscape Gardener.

ELLIOTT'S AMERICAN FRUIT GROWER'S GUIDE IN ORCHARD AND GARDEN; - - - - - - - 1 25

Being a Compend of the History, Modes of Propagation. Culture. &c., of Fruit Trees and Shrubs, with descriptions of nearly all the varieties of Fruits cultivated in this country; and Notes of their adaptation to localities, soils, and a complete list of Fruits worthy of cultivation. By F. R. Elliott, Pomologist.

PRACTICAL FRUIT, FLOWER, AND KITCHEN GARDENER'S COMPANION; - - - - - - - 1 00

With a Calendar. By Patrick Neill, LL.D., F.R.S.E., Secretary of the Royal Caledonian Horticultural Society. Adapted to the United States from the fourth edition, revised and improved by the author. Edited by G. Emerson, M D., Editor of "The American Farmer's Encyclopedia." With Notes and Additions by R. G. Pardee, author of "Manual of the Strawberry Culture." With illustrations

STEPHENS' (HENRY) BOOK OF THE FARM; - - 4 00

A Complete Guide to the Farmer, Steward, Plowman, Cattleman, Shepherd. Field Worker, and Dairy Maid. By Henry Stephens. With Four Hundred and Fifty Illustrations; to which are added Explanatory Notes, Remarks, &c., by J. S. Skinner. Really one of the best books a farmer can possess.

PEDDERS' (JAMES) FARMERS' LAND MEASURER; - 50

Or, Pocket Companion; Showing at one view the Contents of any Piece of Land from Dimensions taken in Yards. With a set of Useful Agricultural Tables.

WHITE'S (W. N.) GARDENING FOR THE SOUTH; - - 1 25

Or, the Kitchen and Fruit Garden, with the best methods for their Cultivation; together with hints upon Landscape and Flower Gardening: containing modes of culture and descriptions of the species and varieties of the Culinary Vegetables. Fruit Trees, and Fruits, and a select list of Ornamental Trees and Plants, found by trial adapted to the States of the Union south of Pennsylvania, with Gardening Calendars for the same. By Wm. N. White, of Athens, Georgia.

EASTWOOD (B.) ON THE CULTIVATION OF THE CRANBERRY; 50

With a Description of the Best Varieties. By B. Eastwood, "Septimus" of the New York Tribune.

AMERICAN BEE-KEEPER'S MANUAL; - - - - 1 00

Being a Practical Treatise on the History and Domestic Economy of the Honey Bee, embracing a full illustration of the whole subject, with the most approved methods of managing this Insect, through every branch of its Culture; the result of many years' experience. Illustrated with many engravings By T. B. Miner.

THAER'S (ALBERT D.) AGRICULTURE   - - - $2 00

The Principles of Agriculture, by Albert D. Thaer; translated by William Shaw and Cuthbert W. Johnson, Esq., F.R.S. With a Memoir of the Author. 1 vol 8vo.

This work is regarded by those who are competent to judge as one of the most beautiful works that has ever appeared on the subject of Agriculture. At the same time that it is eminently practical, it is philoso, bical, and, even to the general reader, remarkably entertaining.

BOUSSINGAULT'S (J. B.) RURAL ECONOMY,   - -   1 25

In its Relations to Chemistry, Physics, and Meteorology: or, Chemistry applied to Agriculture. By J. B. Boussingault. Translated, with notes, etc., by George Law, Agriculturist.

"The work is the fruit of a long life of study and experiment, and its perusal will aid the farmer greatly in obtaining a practical and scientific knowledge of his profession."

MYSTERIES OF BEE-KEEPING EXPLAINED;   - - -   1 00

Being a Complete Analysis of the Whole Subject, consisting of the Natural History of Bees; Directions for obtaining the greatest amount of Pure Surplus Honey with the least possible expense; Remedies for losses given, and the Science of Luck fully illustrated; the result of more than twenty years' experience in extensive apiaries. By M Quinby.

THE COTTAGE AND FARM BEE-KEEPER;   - - -   50

A Practical Work, by a Country Curate.

WEEKS (JOHN M.) ON BEES.—A MANUAL;   - -   50

Or, an Easy Method of Managing Bees in the most profitable manner to their owner; with infallible rules to prevent their destruction by the Moth With an appendix, by Wooster A. Flanders.

THE ROSE;   - - - - - - -   50

Being a Practical Treatise on the Propagation, Cultivation, and Management of the Rose in all Seasons; with a list of Choice and Approved Varieties, adapted to the Climate of the United States; to which is added full directions for the Treatment of the Dahlia. Illustrated by Engravings.

MOORE'S RURAL HAND BOOKS,   - - - -   1 25

First Series, containing Treatises on—

The Horse,	The Pests of the Farm,
The Hog,	Domestic Fowls, and
The Honey Bee,	The Cow,

Second Series, containing—   - - -   1 25

Every Lady her own Flower Gardener,	Essay on Manures,
Elements of Agriculture,	American Kitchen Gardener,
Bird Fancier,	American Rose Culturist.

Third Series, containing—   - -   1 25

Miles on the Horse's Foot,	Vine Dresser's Manual,
The Rabbit Fancier,	Bee-Keeper's Chart,
Weeks on Bees,	Chemistry made Easy.

Fourth Series, containing—   - -   1 25

Persoz on the Vine,	Hooper's Dog and Gun,
Liebig's Familiar Letters,	Skillful Housewife,
	Browne's Memoirs of Indian Corn.

RICHARDSON ON DOGS: THEIR ORIGIN AND VARIETIES.   50

Directions as to their General Management. With numerous original anecdotes. Also. Complete Instructions as to Treatment under Disease. By H D. Richardson. Illustrated with numerous wood engravings.

This is not only a cheap work, but one of the best ever published on the Dog

LIEBIG'S (JUSTUS) FAMILIAR LECTURES ON CHEMISTRY, $0 50

    AND ITS RELATION TO COMMERCE, PHYSIOLOGY, AND AGRICULTURE. Edited by JOHN GARDENER, M.D.

BEMENT'S (C. N.) RABBIT FANCIER;   - - - -   50

    A TREATISE ON THE BREEDING, REARING, FEEDING, AND GENERAL Management of Rabbits, with remarks upon their diseases and remedies, to which are added full directions for the construction of Hutches, Rabbitries, &c., together with recipes for cooking and dressing for the Table. Beautifully illustrated.

THOMPSON (R. D.) ON THE FOOD OF ANIMALS   - -   75

    EXPERIMENTAL RESEARCHES ON THE FOOD OF ANIMALS AND THE Fattening of Cattle; with remarks on the Food of Man. Based upon Experiments undertaken by order of the British Government, by ROBERT DUNDAS THOMSON, M D., Lecturer on Practical Chemistry, University of Glasgow.

THE WESTERN FRUIT BOOK;   - - - -   1 25

    BEING A COMPEND OF THE HISTORY, MODES OF PROPAGATION, CULture, &c., of Fruit Trees and Shrubs, &c., &c. By F. R. ELLIOTT.

THE SKILLFUL HOUSEWIFE;   - - - -   50

    OR COMPLETE GUIDE TO DOMESTIC COOKERY, TASTE, COMFORT, AND Economy, embracing 659 recipes pertaining to Household Duties, the care of Health, Gardening, Birds, Education of Children, &c., &c. By Mrs L. G. ABELL.

THE AMERICAN FLORIST'S GUIDE;   - - -   75

    COMPRISING THE AMERICAN ROSE CULTURIST AND EVERY LADY HER own Flower Gardener.

EVERY LADY HER OWN FLOWER GARDENER;   - -   50

    ADDRESSED TO THE INDUSTRIOUS AND ECONOMICAL ONLY; containing simple and practical Directions for Cultivating Plants and Flowers; also, Hints for the Management of Flowers in Rooms, with brief Botanical Descriptions of Plants and Flowers. The whole in plain and simple language. By LOUISA JOHNSON.

FISH CULTURE;   - - - - - - -   1 00

    A TREATISE ON THE ARTIFICIAL PROPAGATION OF CERTAIN KINDS OF Fish, with the description and habits of such kinds as are most suitable for pisciculture. Also directions for the most successful methods of Angling, illustrated with numerous engravings By THEODATUS GARLICK, M.D., Vice President of Cleveland Academy of Natural Science.

FLINT ON GRASSES;   - - - - - -   1 25

    A PRACTICAL TREATISE ON GRASSES AND FORAGE PLANTS, COMPRISing their natural history, comparative nutritive value, methods of cultivating, cutting, and curing, and the management of grass lands. By CHAS. L. FLINT, A.M., Secretary of Mass. State Board of Agriculture.

WARDER ON HEDGE AND EVERGREENS;   - -   1 00

    A MANUAL ON LIVE FENCES, WITH PARTICULAR DIRECTIONS FOR THEIR planting, culture and trimming, especially with regard to the Maclura hedges, and how to make it. Also an essay on Evergreens, their varieties, propagation, transplanting and culture in the United States. By JOHN A. WARDER, M.D., President of Cincinnati Horticultural Society.

MOORE'S

Hand Books of Rural and Domestic Economy.

All arranged and adapted to the Use of American Farmers.

PRICE 25 CENTS EACH.

HOGS;

THEIR ORIGIN, VARIETIES AND MANAGEMENT. with a View to Profit, and Treatment under Disease: also Plain Directions relative to the most approved modes of preserving their Flesh. By H. D. RICHARDSON, author of "The Hive and the Honey Bee," &c., &c. With Illustrations—12mo.

THE HIVE AND THE HONEY BEE;

WITH PLAIN DIRECTIONS FOR OBTAINING A CONSIDERABLE ANNUAL Income from this branch of Rural Economy: also an Account of the Diseases of Bees and their Remedies, and Remarks as to their Enemies, and the best mode of protecting the Hives from their attacks. By H. D. RICHARDSON. With Illustrations.

DOMESTIC FOWLS;

THEIR NATURAL HISTORY, BREEDING, REARING, AND GENERAL Management. By H. D. RICHARDSON, author of "The Natural History of the Fossil Deer," &c. With Illustrations.

THE HORSE;

THEIR ORIGIN AND VARIETIES; WITH PLAIN DIRECTIONS AS TO THE Breeding, Rearing, and General Management, with Instructions as to the Treatment of Disease. Handsomely Illustrated—12mo. By H. D. RICHARDSON.

THE ROSE;

THE AMERICAN ROSE CULTURIST; being a Practical Treatise on the Propagation, Cultivation, and Management in all Seasons, &c. With full directions for the Treatment of the Dahlia.

THE PESTS OF THE FARM;

WITH INSTRUCTIONS FOR THEIR EXTIRPATION; being a Manual of Plain Directions for the certain Destruction of every description of Vermin. With numerous Illustrations on Wood.

AN ESSAY ON MANURES;

SUBMITTED TO THE TRUSTEES OF THE MASSACHUSETTS SOCIETY FOR Promoting Agriculture, for their Premium. By SAMUEL L. DANA.

THE AMERICAN BIRD FANCIER;

CONSIDERED WITH REFERENCE TO THE BREEDING, REARING, FEEDING, Management, and Peculiarities of Cage and House Birds. Illustrated with Engravings. By D. JAY BROWNE.

CHEMISTRY MADE EASY;

FOR THE USE OF FARMERS. By J. TOPHAM.

ELEMENTS OF AGRICULTURE;

TRANSLATED FROM THE FRENCH, and Adapted to the use of American Farmers. By F. G. SKINNER.

THE HORSE'S FOOT AND HOW TO KEEP IT SOUND;

With Cuts, illustrating the Anatomy of the Foot, and containing valuable hints on shoeing and stable management, both in health and disease. By WILLIAM MILES.

THE SKILLFUL HOUSEWIFE;

Or, Complete Guide to Domestic Cookery, Taste, Comfort, and Economy, embracing 659 recipes pertaining to Household Duties, the care of Health, Gardening, Birds, Education of Children, &c., &c. By Mrs. L. G. ABELL.

THE AMERICAN KITCHEN GARDENER;

CONTAINING DIRECTIONS FOR THE CULTIVATION OF VEGETABLES and Garden Fruits. By T. G. FESSENDEN.

CHINESE SUGAR CANE AND SUGAR MAKING;

ITS HISTORY, CULTURE, AND ADAPTATION TO THE SOIL, CLIMATE, and Economy of the United States, with an account of various processes of Manufacturing Sugar. Drawn from authentic sources by CHARLES F. STANSBURY, A.M., late Commissioner at the Exhibition of all Nations at London.

PERSOZ' CULTURE OF THE VINE;

A NEW PROCESS FOR THE CULTURE OF THE VINE, by PERSOZ. Professor to the Faculty of Sciences of Strasbourg; directing Professor of the School of Pharmacy of the same city. Translated by J. O'C. BARCLAY, Surgeon U. S. N.

THE BEE KEEPER'S CHART;

BEING A BRIEF PRACTICAL TREATISE ON THE INSTINCT, HABITS, and Management of the Honey Bee, in all its various branches, the result of many years' practical experience, whereby the author has been enabled to divest the subject of much that has been considered mysterious and difficult to overcome, and render it more sure, profitable, and interesting to every one, than it has heretofore been. By E. W. PHELPS.

EVERY LADY HER OWN FLOWER GARDENER;

ADDRESSED TO THE INDUSTRIOUS AND ECONOMICAL ONLY; containing Simple and Practical Directions for Cultivating Plants and Flowers; also, Hints for the Management of Flowers in Rooms, with brief Botanical Descriptions of Plants and Flowers. The whole in plain and simple language. By LOUISA JOHNSON.

THE COW; DAIRY HUSBANDRY AND CATTLE BREEDING.

By M. M. MILBURN, and revised by H. D. RICHARDSON and AMBROSE STEVENS. With Illustrations.

WILSON ON THE CULTURE OF FLAX;

ITS TREATMENT, AGRICULTURAL AND TECHNICAL; delivered before the New York State Agricultural Society, at the Annual Fair at Saratoga, in September last, by JOHN WILSON, late President of the Royal Agricultural College at Cirencester, England.

WEEKS ON BEES: A MANUAL.

OR, AN EASY METHOD OF MANAGING BEES IN THE MOST PROFITABLE manner to their owner, with infallible rules to prevent their destruction by the Moth; with an Appendix by WOOSTER A. FLANDERS.

REEMELIN'S (CHAS.) VINE DRESSER'S MANUAL;

CONTAINING FULL INSTRUCTIONS as to LOCATION and SOIL; Preparation of Ground; Selection and Propagation of Vines; the Treatment of a Young Vineyard; trimming and training the vines; manures and the making of wine. Every department illustrated.

HYDE'S CHINESE SUGAR CANE;

CONTAINING ITS HISTORY, MODE OF CULTURE, MANUFACTURE of the Sugar, &c.; with Reports of its success in different parts of the United States.

BEMENT'S (C. M.) RABBIT FANCIER ;

A TREATISE ON THE BREEDING, REARING, FEEDING, AND GENERAL Management of Rabbits, with remarks upon their diseases and remedies; to which are added full directions for the construction of Hutches, Rabbitries, &c., together with recipes for cooking and dressing for the table.

RICHARDSON ON DOGS: THEIR ORIGIN AND VARIETIES ;

DIRECTIONS AS TO THEIR GENERAL MANAGEMENT.   With numerous original anecdotes. Also Complete Instructions as to Treatment under Disease. By H. D. RICHARDSON. Illustrated with numerous wood engravings.
This is not only a cheap, but one of the best works ever published on the Dog.

LIEBIG'S (JUSTUS) FAMILIAR LETTERS ON CHEMISTRY ;

AND ITS RELATION TO COMMERCE, PHYSIOLOGY, AND AGRICULTURE. Edited by JOHN GARDENER, M. D.

THE DOG AND GUN ;

A FEW LOOSE CHAPTERS ON SHOOTING, among which will be found some Anecdotes and Incidents.   Also Instructions for Dog Breaking, and interesting letters from Sportsmen. BY A BAD SHOT.

THE PRESERVATION OF FOOD ;

THE VARIOUS METHODS OF PRESERVING MEATS, FRUITS, VEGETABLES, Milk, Butter, Grain, &c., by drying, smoking, pickling, and other processes. By E. GOODRICH SMITH.